Analysis of Boundedness and Safeness in a Petri Net-Based Specification of Concurrent Control Systems

Authored by

Marcin Wojnakowski

Institute of Control & Computation Engineering
University of Zielona Góra,
ul. Szafrana 2, 65-516 Zielona Góra, Poland

Analysis of Boundedness and Safeness in a Petri Net-Based Specification of Concurrent Control Systems

Author: Marcin Wojnakowski

ISBN (Online): 978-981-5322-70-5

ISBN (Print): 978-981-5322-71-2

ISBN (Paperback): 978-981-5322-72-9

First published in 2025.

Bentham Science Publishers Pte. Ltd.
80 Robinson Road #02-00
Singapore 068898
Singapore
Email: subscriptions@benthamscience.net

BENTHAM SCIENCE

CONTENTS

FOREWORD i
PREFACE iii
DEDICATION iv
ACKNOWLEDGEMENT v

CHAPTER 1 CONTROL SYSTEMS 1
1.1 INTRODUCTION 1
1.2 CHALLENGES 3
1.3 MAIN PURPOSES 4
CONCLUSION 5
REFERENCES 5

CHAPTER 2 THEORETICAL ASPECTS OF PETRI NETS 10
2.1 INTRODUCTION 10
2.2 ALGORITHMS 10
2.3 COMPUTATIONAL COMPLEXITY 11
2.4 GRAPHS 12
2.5 PETRI NET-BASED SPECIFICATION 13
2.6 BASIC NOTATIONS AND DEFINITIONS 14
2.7 CLASSES OF PETRI NETS 23
2.7.1 State Machines 24
2.7.2 Marked Graphs 25
2.7.3 Other Classes 26
2.8 REDUCED ROW ECHELON FORMS 27
CONCLUSION 30
REFERENCES 30

CHAPTER 3 BOUNDEDNESS AND SAFENESS 37
3.1 INTRODUCTION 37
3.2 COVERABILITY GRAPHS 38
3.2.1 Overview 39
3.2.2 Construction 40
3.3 INVARIANTS 41
3.4 PROPERTY OF BOUNDEDNESS 45
3.5 PROPERTY OF SAFENESS 49
CONCLUSION 53
REFERENCES 53

CHAPTER 4 METHODS OF ANALYZING BOUNDEDNESS AND SAFENESS 62
4.1 BOUNDEDNESS 62
4.1.1 A Method Based on Reachability Graphs 62
4.1.2 A Method Based on Invariants 65
4.1.3 A Method Based on Reduced Row Echelon Form 69
4.2 SAFENESS 75
4.2.1 A Method Based on Reachability Graphs 75
4.2.2 A Method Based on Invariants 75
CONCLUSION 80
REFERENCES 80

CHAPTER 5 EXPERIMENTAL VERIFICATION OF THE PROPOSED METHODS 82
5.1 INTRODUCTION 82
5.2 THE HIPPO SYSTEM 83
5.3 BOUNDEDNESS 84
5.4 SAFENESS 90
CONCLUSION 94
REFERENCES 94

CHAPTER 6 EFFECTIVE AND EFFICIENT ANALYSIS OF BOUNDEDNESS AND SAFENESS - CASE STUDY 95
6.1 INTRODUCTION 95
6.2 CASE STUDY EXAMPLE 96
6.3 BOUNDEDNESS AND SAFENESS ANALYSIS 97
6.4 CONFIRMATION OF THE EFFECTIVE AND EFFICIENT ANALYSIS 101
6.5 CONTRIBUTION SUMMARY 101
CONCLUSION 102
REFERENCES 103
APPENDIX A DETAILED EXPERIMENTAL RESULTS 105
A.1 BOUNDEDNESS 105
A.2 SAFENESS 132
ACRONYMS 146
SUBJECT INDEX 148

FOREWORD

I am very pleased to introduce here a comprehensive and well-balanced book on the analysis of boundedness and safeness in Petri net-based specification of discreteevent concurrent control systems.

In the dynamic landscape of control systems, Petri nets emerge as a powerful modeling approach. Their simplicity, combined with their ability to represent complex systems, has made them essential in various domains. Petri nets provide a robust framework for specification and analysis of systems in very diverse fields, such as software engineering, robotics and automation, manufacturing and industrial engineering, telecommunications, transportation and logistics, and healthcare and medical systems, just to mention a few.

I particularly like the structure of the book. The first chapter covers introductory topics on control systems and Petri nets, followed by three chapters addressing boundedness and safeness in Petri nets models, where new methods are presented.

The last two chapters cover two case studies, complemented by an appendix containing extensive results of the application of the proposed methods for verification of properties of boundedness and safeness.

The journey with this book begins with a comprehensive overview of the tools and algorithms necessary for analyzing the boundedness and safeness of Petri nets. The novel algorithms proposed in this work are not just theoretical constructs; they are rigorously validated through extensive experimentation on a varied set of 243 Petri nets. This empirical approach provides valuable insights into the practical applicability and limitations of these algorithms. To illustrate the real-world impact, this publication presents a manufacturing control system modeled using Petri nets.

Through this example, the author serves a concrete demonstration of how theoretical principles can be effectively translated into practical solutions, offering a guide for both researchers and industry practitioners.

I am confident that "Analysis of Boundedness and Safeness in a Petri Net Based Specification of Concurrent Control Systems" will become an important resource for researchers, educators, and professionals involved in the design, analysis, and implementation of concurrent control systems based on Petri nets. Its synthesis of theoretical rigor and practical relevance makes it a standout contribution to the field of computer science and engineering.

Summing up my overall impression about this book, I strongly recommend it to anyone who likes to use Petri net models in systems design, emphasizing boundedness and safeness properties verification. The book offers insightful contributions, and may be used as a text for advanced courses on the subject.

Luís Gomes
Department of Electrical and Computer Engineering,
NOVA University Lisbon,
Lisbon, Portugal.

PREFACE

Control systems play a crucial role in various aspects of everyday life, including banking, medical care, manufacturing, transportation, and entertainment. Their continual development necessitates designers to employ advanced and multi-functional tools to facilitate the design process. Petri nets stand out as an effective modeling approach in this context. They have gained popularity for their ease of analysis and graphical representation. Consequently, the analysis of boundedness and safeness of systems specified by Petri nets poses a significant challenge. This book offers a comprehensive overview of applications and algorithms for analyzing the boundedness and safeness of Petri nets.

Boundedness indicates a finite number of reachable states in a control system, while safeness characterizes a binary behavior essential for Petri nets used in configurable FPGAs. Given the exponential computational complexity of Petri net properties, the book proposes novel algorithms to address this analytical gap. These new solutions are elaborately described and supported by experimental results from a set of 243 Petri nets, along with a discussion of their limitations. Furthermore, a real-life manufacturing control system modeled with a Petri net is presented to highlight the benefits of designing concurrent systems using bounded and safe nets.

Marcin Wojnakowski
Institute of Control & Computation Engineering,
University of Zielona Góra,
ul. Szafrana 2, 65-516 Zielona Góra, Poland.

DEDICATION

I dedicate the book to my beloved wife and to bl. Carlo Acutis, who is the patron of this work.

ACKNOWLEDGEMENT

I would like to express my deep gratitude to Professor Remigiusz Wiśniewski for involving me in fascinating research procedures and publication of many scientific papers and this book.

I would like to extend my thanks to my colleagues Dr. Grzegorz Bazydło and Mateusz Popławski, who provided valuable advice regarding my research. I also want to thank Dariusz Nowosad from the University Language Center of University of Zielona Góra for his professional language proofreading.

Finally, I wish to thank my wife and parents for their support.

CHAPTER 1

Control Systems

Abstract: Control systems surround people everywhere. Recent years have witnessed wide development of such systems. This is due to the growing demand for systems that autonomously control our environment. For example, there are lamps that automatically turn on at dusk, heating systems that turn offwhen we go to work, and similar ideas that make our lives easier, cheaper, and more ecological. These so-called smart solutions can be found in various aspects of human life, such as banking, manufacturing, automotive, transportation, medical care, process mining, and others.

Keywords: Analysis, Boundedness, Concurrent control system, Cyber-physical system, Embedded system, Petri net, Safeness.

1.1 INTRODUCTION

Control systems can be seen as systems that manage other systems, known as operational systems [1]. Such systems send appropriate signals to operational systems based on input values and logic conditions [2]. Several years ago, there were simple control systems with few input and output signals. In recent years, designers have elaborated complex, concurrent, and advanced systems, commonly known as embedded systems, Internet of Things (IoT), flexible manufacturing systems (FMS), and cyber-physical systems (CPS), which is the most popular nowadays [3-34]. Since it is impossible to design such sophisticated systems without the application of computer-aided design (CAD) tools, CAD software is used, for *e.g.*, to perform modeling based on finite state machines, UML diagrams, Petri nets, interpreted nets, *etc.* [11].

Example 1.1 (Control System) : Let us imagine a real-life situation and assume that there is a moving car near a playground where children are playing. One child suddenly crosses the road in front of the car. Now consider a modern car with advanced

Marcin Wojnakowski

control systems on board. In such a case, the car's front sensors simultaneously detect an object moving into the road. A dedicated control system simultaneously analyzes the input data from the front sensors and decides to send an emergency signal to the operational system. As a result, the car is automatically halted. Such solutions are offered by Bosch Mobility Solutions[1] (Fig. **1.1**) or Continental-Automotive[2] (Fig. **1.2**). In this way, various control systems make our lives easier, cheaper, and also safer.

Fig. (1.1). Automatic emergency braking.

---

1 More information available online: https://www.bosch-mobility-solutions.com/en/solutions/assistance-systems/automatic-emergency-braking (accessed on Novem- ber 30, 2024)

2 More information available online: https://www.continental-automotive.com/en-gl/Passenger- Cars/Autonomous-Mobility/Functions/Cruising-Driving/Emergency-Brake-Assist/Pedestrian (accessed on November 30, 2024)

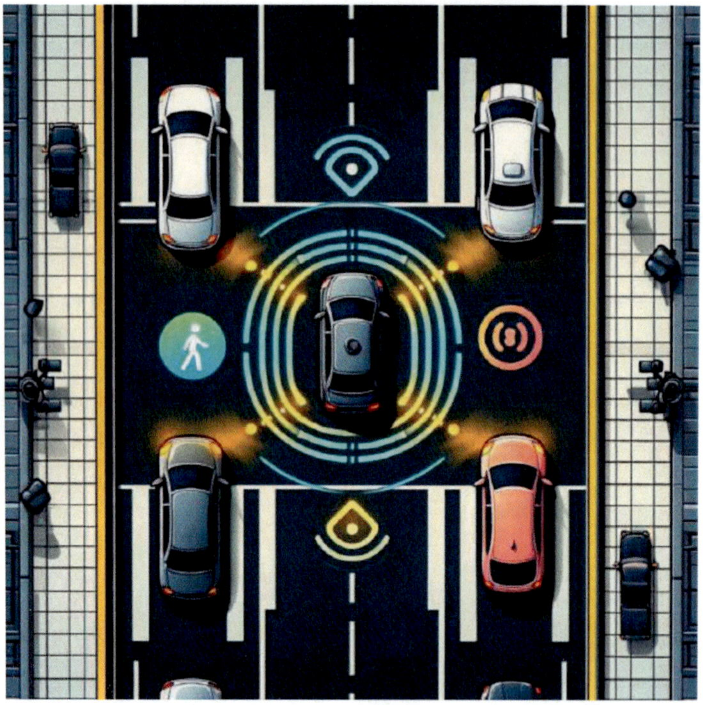

Fig. (1.2). Emergency brake assist – pedestrian.

1.2 CHALLENGES

The real-life automotive example attests to dynamic development and increasing levels of complexity in modern concurrent control systems. It encourages the use of advanced modeling (specifying) approaches, one of which is Petri nets, which, in addition to graphic (graph-based) representation, allow for formal description (specification) [1, 35]. Moreover, Petri nets are also supported by formal mechanisms that make it possible to analyze models. Such analyses permit the examination of systems' reliability and robustness at the specification stage, which may impact the time and costs of a designed concurrent control system [9].

In their very recent research, various scientists have proved the profitability and usefulness of the idea of Petri nets, for *e.g.*, Gomes *et al.* [36], Grobelna *et al.* [9], Lee and Seshia [35], Nuño-Sánchez *et al.* [37], Ramírez-Trevino *et al.* [38], and Wiśniewski *et al.* [1]. Nevertheless, their research also gives rise to an analytical problem regarding the analysis of Petri nets used for concurrent control system specifications [1, 9, 11, 39-42]. Such analyses of Petri net properties can be performed based on state space exploration or by means of a linear Algebraic approach (solving linear equations) [39, 40, 43]. The former approach examines a state explosion problem, which means that as systems grow larger, the number of all possible states grows exponentially [44, 45]. In the linear algebraic approach, the number of computed invariants can also grow exponentially [1, 40, 44]. Hence, both approaches are limited by exponential complexity in relation to the time they need to produce results in the general case [46]. This way, results cannot always be obtained in the assumed time. In such a case, we conclude that the selected method is not efficient [44].

Boundedness and safeness are the most important properties of Petri net-based specifications [39, 40]. Boundedness is responsible for the finite number of reachable states of a control system [39, 47]. In other words, designers expect systems whose behavior can be controlled and all their states can be determined. Therefore, there should be no states of the system that cannot be predicted in the specification. Hence, the Petri nets used in formal specifications of concurrent control systems are required to be bounded [39, 40, 44]. In addition, there is sometimes a need to use safe (*stronger* boundedness) Petri nets [40, 47]. Safe Petri nets are capable of logical control, *i.e.*, they are able to indicate whether an action should be active or inactive (on/ off nature) [1].

1.3 MAIN PURPOSES

The following proposition has been formulated for this work:

Proposition 1.1 : The analysis of boundedness and safeness of concurrent control systems specified by Petri nets can be performed effectively and efficiently.

Furthermore, the book has the following purposes:

- to conduct a broad survey on the boundedness and safeness of Petri net properties
- to present novel algorithms for analyzing boundedness
- to present novel algorithms for analyzing safeness
- to verify effectiveness and efficiency of the recent methods.

According to the classical definition, the efficiency of an algorithm means that the results can be found within the assumed time [44]. In this work, the assumed time is fixed to one hour. Furthermore, the effectiveness of an algorithm means that the results obtained are correct [44]. In other words, the results are consistent with the definition of the property under test.

CONCLUSION

The first chapter of this book includes introduction to the subject matter, formulation of its objective, and motivation behind its creation. Basic concepts of the theory of Petri nets are introduced in the second chapter, which features a discussion of their most important properties, such as boundedness and safeness. An overview of the literature and available analysis methods are presented in the third chapter, together with a formulation of problems related to property verification. The fourth chapter introduces novel ideas for procedural analysis and presents preliminaries that are indispensable in understanding the ideas. Contrary to widely recognized algorithms under some assumptions, recent solutions are efficient. Moreover, the novel methods are applied to a real-life large-scale system, while the knowable algorithms are more theoretical ideas. Subsequently, experimental results that focus on efficiency and effectiveness are presented in the fifth chapter. When the theory of Petri nets and the innovative analysis solutions are explicated, a case-study of manufacturing control specification based on Petri net is introduced in the sixth chapter. Finally, the last chapter summarizes the book and confirms its proposition.

REFERENCES

[1] R. Wiśniewski, *Prototyping of Concurrent Control Systems Implemented in FPGA Devices*. Advances in Industrial Control, Springer International Publishing, 2017.

[2] M. Wiśniewska, *Application of hypergraphs in decomposition of discrete systems. Lecture Notes in Control and Computer Science*. Zielona Góra: University of Zielona Góra Press, vol. 23, 2012.

[3] E. A. Lee, "Cyber physical systems: Design challenges", in *2008 11th IEEE International Symposium on Object and Component-Oriented Real-Time Distributed Comput- ing (ISORC)*, pp. 363–369, 2008. ISSN: 2375-5261.

[4] J. Lee, B. Bagheri, and H.-A. Kao, "A cyber-physical systems architecture for industry 4.0-based manufacturing systems," *Manufac. Let.*, vol. 3, pp. 18–23, 2015.

[5] P.-A. Hsiung, T.-W. Kuo, Y.-H. Chang, and C.-H. Huang, "Introduction to the special issue on reconfigurable cyber-physical and embedded system design," *J. of Sys. Arch.*, vol. 62, p. 39, 2016.

[6] Y. Wang, G. Tan, Y. Wang, and Y. Yin, "Perceptual control architecture for cyber–physical systems in traffic incident management," *J. of Sys. Arch.*, vol. 58, pp. 398–411, 2012.

[7] M. Abdel-Basset, G. Manogaran, A. Gamal, and F. Smarandache, "A hybrid approach of neutrosophic sets and DEMATEL method for developing supplier selection criteria," *Des Autom Embed Syst*, vol. 22, pp. 257–278, 2018.

[8] R. Wiśniewski, M. Wiśniewska, and M. Jarnut, "C-exact hypergraphs in concurrency and sequentiality analyses of cyber-physical systems specified by safe petri nets," *IEEE Access*, vol. 7, pp. 13510–13522, 2019.

[9] I. Grobelna, R. Wiśniewski, and M. Wojnakowski, "Specification of Cyber-Physical Systems with the Application of Interpreted Nets," in *IECON 2019 - 45th Annual Conference of the IEEE Industrial Electronics Society*, vol. 1, pp. 5887–5891, 2019. ISSN: 2577-1647.

[10] R. Wiśniewski, G. Bazydło, P. Szcześniak, I. Grobelna, and M. Wojnakowski, "Design and verification of cyber-physical systems specified by petri nets—A case study of a direct matrix converter," *Mathematics*, vol. 7, p. 812, 2019.

[11] R. Wiśniewski, G. Bazydło, L. Gomes, A. Costa, and M. Wojnakowski, "Analysis and design automation of cyber-physical system with hippo and iopt-tools," in *IECON 2019 - 45th Annual Conference of the IEEE Industrial Electronics Society*, vol. 1, pp. 5843–5848, 2019. ISSN: 2577-1647.

[12] J. Suk, Y. Lee, S. Kim, H. Koo, and J. Kim, "System identification and stability evaluation of an unmanned aerial vehicle from automated flight tests," *KSME Inter. J.*, vol. 17, pp. 654–667, 2003.

[13] S. S. Peng and M. C. Zhou, "Ladder diagram and Petri-net-based discrete-event control design methods," *IEEE transactions on systems, man, and cybernetics, part c (applications and reviews)*, vol. 34, pp. 523–531, 2004.

[14] L. Gomes, J. P. Barros, and A. Costa, "Modeling formalisms for embedded system design, embedded systems handbook," in *Embedded Systems Handbook*, Taylor and Francis Group, LLC, 2006.

[15] A. Attoui, "An environment based on rewriting logic for parallel systems formal spec- ification and prototyping," *J. of Sys. Arch.*, vol. 44, pp. 79–105, 1997.

[16] L. Dai and W. Guo, "Concurrent subsystem-component development model (CSCDM) for developing adaptive e-commerce systems," in *Computational Science and Its Applications – ICCSA 2007* (O. Gervasi and M. L. Gavrilova, Eds.), Lecture Notes in Computer Science, (Berlin, Heidelberg), pp. 81–91, Springer, 2007.

[17] A. Santone, V. Intilangelo, and D. Raucci, "Application of equivalence checking in a loan origination process in banking industry," in *2013 Workshops on Enabling Technologies: Infrastructure for Collaborative Enterprises*, pp. 292–297, 2013. ISSN: 1524-4547.

[18] S. Kirn, R. Unland, and U. Wanka, "MAMBA: Automatic customization of computerized business processes," *Infor. Sys.*, vol. 19, pp. 661–682, 1994.

[19] P. S. Andrews and J. Timmis, "Inspiration for the Next Generation of Artificial Immune Systems," in *Artificial Immune Systems* C. Jacob, M. L. Pilat, P. J. Bentley, and J. I. Timmis, eds., Lecture Notes in Computer Science, Berlin, Heidelberg, pp. 126–138, Springer, 2005.

[20] M. Chen and R. Hofestädt, "Quantitative petri net model of gene regulated metabolic networks in the cell," *Stud Health Technol Inform*, vol. 162, pp. 38–55, 2011.

[21] K. Chinzei, N. Hata, F. A. Jolesz, and R. Kikinis, "MR compatible surgical assist robot: System integration and preliminary feasibility study," in *Medical Image Computing and Computer-Assisted Intervention – MICCAI 2000* S. L. Delp, A. M. DiGoia, and B. Jaramaz, Eds., Lecture Notes in Computer Science, Berlin, Heidelberg, pp. 921– 930, Springer, 2000.

[22] H. Hu, M. Zhou, and Z. Li, "Supervisor design to enforce production ratio and absence of deadlock in automated manufacturing systems," *IEEE Trans. on Sys., Man, and Cyber. - Part A: Sys. and Hum.*, vol. 41, pp. 201–212, 2011.

[23] R. W. Lewis, *Programming Industrial Control Systems Using IEC 1131-3*. IET Digital Library, 1998.

[24] Y. Chen, Z. Li, and A. Al-Ahmari, "Nonpure petri net supervisors for optimal deadlock control of flexible manufacturing systems," *IEEE Tran. on Sys., Man, and Cyber.: Sys.*, vol. 43, pp. 252–265, 2013.

[25] Z. Li, M. Zhou, and N. Wu, "A survey and comparison of petri net-based deadlock prevention policies for flexible manufacturing systems," *IEEE Trans. on Sys., Man, and Cyber., Part C (Applic. and Rev.),* vol. 38, pp. 173–188, 2008.

[26] M. Alcaraz-Mejia, E. Lopez-Mellado, and A. Ramírez-Trevino, "Redundancy based controller reconfiguration for fault recovery of manufacturing systems," in *2007 IEEE International Conference on Automation Science and Engineering*, pp. 128–133, 2007. ISSN: 2161-8089.

[27] M. Abdel-Basset, G. Manogaran, L. Abdel-Fatah, and S. Mirjalili, "An improved nature inspired meta-heuristic algorithm for 1-D bin packing problems," *Pers Ubiquit Comput*, vol. 22, pp. 1117–1132, 2018.

[28] M. Abdel-Basset, G. Manogaran, H. Rashad, and A. N. H. Zaied, "A comprehensive review of quadratic assignment problem: variants, hybrids and applications," *J Ambient Intell Human Comput*, 2018.

[29] J. Carmona, J. Cortadella, and M. Kishinevsky, "Divide-and-conquer strategies for process mining," in *Business Process Management,* (U. Dayal, J. Eder, J. Koehler, and H. A. Reijers, Eds. Lecture Notes in Computer Science, Berlin, Heidelberg, pp. 327– 343, Springer, 2009.

[30] S. Goedertier, D. Martens, J. Vanthienen, and B. Baesens, "Robust process discovery with artificial negative events," *J. Mach. Learn. Res.*, vol. 10, pp. 1305–1340, 2009.

[31] N. A. Lynch, "Multilevel atomicity—a new correctness criterion for database concur- rency control," *ACM Trans. Database Syst.*, vol. 8, pp. 484–502, 1983.

[32] W. M. P. v. d. Aalst, "Decomposing Petri nets for process mining: A generic approach," *Distrib. and Paral. Data.*, 2013.

[33] V. Gourcuff, O. De Smet, and J.-M. Faure, "Efficient representation for formal verifica- tion of PLC programs," in *2006 8th International Workshop on Discrete Event Systems*, pp. 182–187, 2006.

[34] A. Barkalov, L. Titarenko, and J. Bieganowski, "Synthesis of compositional micro-program control unit with extended microinstruction format," in *2009 MIXDES-16th International Conference Mixed Design of Integrated Circuits Systems*, pp. 328–331, 2009.

[35] E. A. Lee and S. A. Seshia, *Introduction to Embedded Systems: A Cyber-Physical Systems Approach.* 2nd ed, Cambridge, Massachuetts: The MIT Press, 2016.

[36] L. Gomes and J. P. Barros, "Refining IOPT petri nets class for embedded system controller modeling," in *IECON 2018 - 44th Annual Conference of the IEEE Industrial Electronics Society*, pp. 4720–4725, 2018. ISSN: 2577-1647.

[37] S.-A. Nuño-Sánchez, A. Ramírez-Treviño, and J. Ruiz-León, "Structural sequence detectability in free choice interpreted petri nets," *IEEE Trans. on Auto. Con.*, vol. 61, pp. 198–203, 2016.

[38] A. Ramírez-Trevino, E. Ruiz-Beltran, I. Rivera-Rangel, and E. Lopez-Mellado, "Online fault diagnosis of discrete event systems. A petri net-based approach," *IEEE Trans. on Auto. Sci. and Engin.*, vol. 4, pp. 31–39, 2007.

[39] M. Wojnakowski and R. Wiśniewski, "Verification of the boundedness property in a petri net-based specification of the control part of cyber-physical systems". In: *Technological Innovation for Applied AI Systems*. DoCEIS 2021. *IFIP Advances in Information and Communication Technology Camarinha-Matos*, L.M., Ferreira, P., Brito, G. (eds), Springer, Cham. vol 626, pp 83-91. 2021.

[40] M. Wojnakowski, R. Wiśniewski, G. Bazydło, and M. Popławski, "Analysis of safeness in a Petri net-based specification of the control part of cyber-physical systems," *AMCS*, vol. 31, pp. 647–657, 2021.

[41] R. Wiśniewski, A. Karatkevich, L. Stefanowicz, and M. Wojnakowski, "Decomposition of distributed edge systems based on the Petri nets and linear algebra technique," *J. of Sys. Arch.*, vol. 96, pp. 20–31, 2019.

[42] R. Wiśniewski, G. Bazydło, P. Szcześniak, and M. Wojnakowski, "Petri net-based specification of cyber-physical systems oriented to control direct matrix converters with space vector modulation," *IEEE Access*, vol. 7, pp. 23407–23420, 2019.

[43] M. Wojnakowski, M. Popławski, R. Wiśniewski, and G. Bazydło, "Safeness analysis of petri net-based cyber-physical systems based on the linear algebra and parallel reductions," in *AIP Conference Proceedings*, Heraklion, Greece, 2021.

[44] M. Wojnakowski, M. Popławski, R. Wiśniewski, and G. Bazydło, "Hippo-CPS: verification of boundedness, safeness and liveness of petri net-based cyber-physical systems," *in Technological Innovation for Digitalization and Virtualization, IFIP Adv. Inf. Commun. Technol.* L. M. Camarinha-Matos, Ed., Cham: Springer Int. Publ., pp. 74–82, 2022.

[45] M. Popławski, M. Wojnakowski, G. Bazydło, and R. Wiśniewski, "Reachability tree in liveness analysis of petri net-based cyber-physical systems," *in AIP Conference Proceedings*, Heraklion, Greece, 2021.

[46] R. J. Lipton, "The reachability problem requires exponential space," in *Technical Report 62*, Yale University, 1976.

[47] A. Karatkevich, *Dynamic Analysis of Petri Net-Based Discrete Systems*. Lecture Notes in Control and Information Sciences, Berlin Heidelberg: Springer-Verlag, 2007.

CHAPTER 2

Theoretical Aspects of Petri Nets

Abstract: A Petri net is one of the forms of graphical specification and representation of various concurrent control systems. Wide applications of Petri nets can be found in the field of distributed systems, embedded systems, edge computing, manufacturing systems, and cyber-physical systems. They are a useful modeling approach because they are supported by verification, validation, and analytical methods. This way, system designers are able to verify the robustness and reliability of their projected systems.

Keywords: Algorithm, Boundedness, Computational complexity, Concurrent control system, Directed graph, Petri net-based specification, Safeness, Verification.

2.1 INTRODUCTION

This chapter opens with a discussion on the computational complexity of algorithms, as algorithms can be seen as procedures that solve computational problems. Subsequently, there are examples of graphs as an introduction to Petri nets. Finally, underlying definitions and notation related to Petri nets and their fundamental properties are presented. Then, theories and definitions are discussed that illustrate in detail the currently available algorithms (supported by simple examples). Chapter 4 addresses novel methods to analyze the properties of boundedness and safeness.

2.2 ALGORITHMS

First of all, the analysis of the properties of Petri nets, such as boundedness and safeness, can be treated as a computational problem to be solved. *An algorithm* can be viewed as a method for solving a well-specified computational problem [1]. Formally, an algorithm is any well-defined computational procedure that takes a set of values as *input* and returns a set of values as *output* [1]. Therefore, it is a sequence of steps that convert input into output. In our considerations, Petri nets constitute the input to algorithms and the output will result from the analysis of a relevant net property.

Marcin Wojnakowski

Example 2.1 (Algorithm) : Consider a simple computational problem of computation of absolute values as an example. Algorithm 2.1 proposes a method to solve the problem. We have the following set of inputs of real numbers: $S_{in} = \{x_1, x_2, x_3, \cdots, x_n\}$ and expect the following set of outputs: $S_{out} = \{|x_1|, |x_2|, |x_3|, \cdots, |x_n|\}$. Firstly, an empty set is assigned to S_{out}. Then, for each number x of the input set, the value is calculated: $|x|$ and added to the output set S_{out}. Finally, the output set contains absolute values of the numbers from the input set.

```
Data: set S_in = {x_1, x_2, x_3, ··· , x_n}
Result: set S_out = {|x_1|, |x_2|, |x_3|, ··· , |x_n|}
S_out ← ∅ ;
foreach variable x of set S_in do
  | add |x| to S_out ;
end
```

Algorithm 2.1: An algorithm example — calculation of the absolute value.

2.3 COMPUTATIONAL COMPLEXITY

Each computer algorithm can be characterized by its computational complexity. Computational complexity is necessary to estimate the efficiency of a method. Below are some preliminaries related to the computational complexity of algorithms [2], which we will describe in the subsequent chapters of this book.

Definition 2.1 *Time complexity* of an algorithm is the function $f : \mathbb{N} \rightarrow \mathbb{N}$ such that $f(n)$ is the maximum number of iterations that the algorithm uses on any input of length n.

Definition 2.2 *An upper bound* for $f(n)$ is $g(n)$ that $f(n) = O(g(n))$.

Definition 2.3 An algorithm is bounded by *a polynomial* in the size of inputs n if the number of its iterations is estimated as

$$f(n) = O(n^c) , \tag{2.1}$$

where $c > 0$.

Definition 2.4 An algorithm is bounded by *an exponential* in the size of inputs n if the number of its iterations is estimated as

$$f(n) = O(c^n) , \tag{2.2}$$

where $c > 1$.

Definition 2.5 *A polynomial complexity (polynomial time)* of an algorithm indicates that the total run time of the algorithm to generate all outputs is bounded by a polynomial in the size of the input.

Definition 2.6 *An exponential complexity (exponential time)* of an algorithm means that the total run time of the algorithm to generate all outputs is bounded by an exponential in the size of inputs.

Example 2.2 (Computational Complexity) : The computational complexity of Algorithm 2.1 can be estimated as follows. The *foreach* loop on lines 2–3 is performed $n = |S_{in}|$ times. Hence, the algorithm is bounded by the function $O(n)$, where n is the number of elements in the input set S_{in}. It is said to be characterized by polynomial (first degree) computational complexity.

> Efficiency and Effectiveness

Computer science algorithms are usually measured in two directions: execution time and memory usage by the program. The *efficiency* term in this book refers to the optimization (minimization) of the execution of the run time with respect to the response in the assumed time. Thus, it does not include other aspects. The focus on improvements is only to reduce running time while maintaining correct results (*effectiveness*).

2.4 GRAPHS

Before we define Petri nets, we shall introduce some basic concepts and notation related to graph theory [3-6].

Definition 2.7 *A graph* is defined by a pair

$$G = (V, \mathbb{E}) , \tag{2.3}$$

where

- $V = \{v_1, v_2, \cdots, v_n\}$ is a finite, nonempty set of vertices,
- $\mathbb{E} = \{E_1, E_2, \cdots, E_m\}$ is a finite set of *unordered* pair of vertices, called edges.

Definition 2.8 *A digraph (directed graph)* is a pair

$$D = (V, \mathbb{E}) , \tag{2.4}$$

where

- $V = \{v_1, v_2, \cdots, v_n\}$ is a finite, nonempty set of vertices,
- $\mathbb{E} = \{E_1, E_2, \cdots, E_m\}$ is a finite set of an *ordered* pair of vertices, called edges.

Example 2.3 (Directed Graph) : The usual way to illustrate a graph is by drawing a circle corresponding to each vertex, joining by a line for an edge and an arrow that informs of the direction (only in case of directed graphs). Fig. (**2.1**) shows a simple digraph that consists of six vertices: $V = \{v_1, v_2, v_3, v_4, v_5, v_6\}$ and seven edges: $E = \{E_1, E_2, E_3, E_4, E_5, E_6, E_7\}$. Each edge forms an ordered pair of vertices: $E_1 = (v_1, v_2)$, $E_2 = (v_2, v_3)$, $E_3 = (v_3, v_4)$, $E_4 = (v_4, v_1)$, $E_5 = (v_4, v_5)$, $E_6 = (v_6, v_4)$, $E_7 = (v_5, v_6)$.

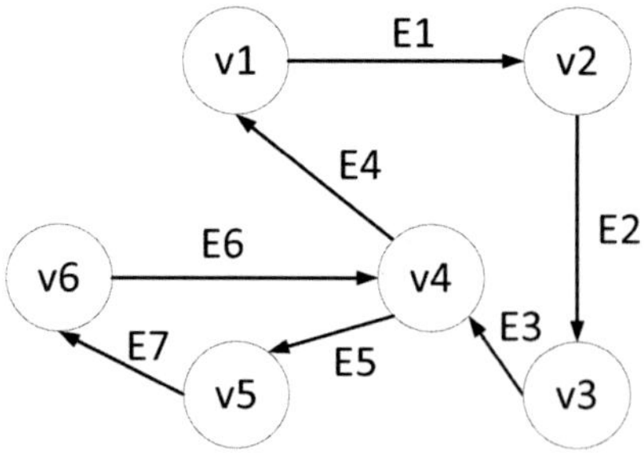

Fig. (2.1). An example of a directed graph.

2.5 PETRI NET-BASED SPECIFICATION

Algorithms in the computer science area can be represented in several ways [7]. The popular ones are based on block diagrams or concurrency flow diagrams supported by mathematical equations [7]. Today, it seems that more capable methods apply dedicated modeling languages, for instance, UML [8, 9] or Petri nets [10, 11]. In this book, we rely on Petri nets because they yield several benefits over the other approaches named here [7, 12-23]. First, the Petri net specifying technique perfectly exploits the parallel operations of a specification [7, 21-27]. It is important since most of the operations in concurrent control systems such as cyber-physical systems [2, 7, 28-36] are performed simultaneously [2, 7, 21, 22].

Furthermore, various verification, validation, and analysis tools supplement Petri nets [13, 14, 26, 37-42]. Methods are used in order to ensure that the formal specification is correct [24, 26, 28, 38, 43-49]. The correctness of the model is provided before going on with the next steps of the design path [7, 32, 50-55]. This advantage is especially significant in industries [37, 56-58] where each return to previous stages is very costly [55]. Sometimes, prototyping of concurrent control systems involves splitting of the system into separate subsystems [7, 21, 22, 59]. Petri nets can be automatically decomposed into a set of modules [59]. Additionally, Pereira *et al.* [60] proposed a comfortable prototyping framework that includes a graphical editor for Petri net modeling, a state space generator (verification), and a manual simulator (validation).

The idea of Petri nets was introduced by Carl Adam Petri in his dissertation in 1962 [4, 61, 62, 63]. A Petri net [64] is a directed graph where vertices are places and edges function as transitions. As a digraph, a Petri net can be converted into another strict mathematical notation [62], such as the incidence matrix [65]. Thus, we can use graph theory and linear algebra for various analyses of Petri nets (system models). It is impossible in other approaches, such as UML design [9], where we do not have a mathematical structure. Petri nets are still being developed and applied to various disciplines of science and industrial uses [27, 64, 66-74].

This makes it necessary to modify and introduce their new types, *e.g.*, hybrid Petri nets [75], timed continuous Petri nets [76], colored Petri nets, ordinary Petri nets, interpreted Petri nets, and unbounded Petri nets. The multitude of types attests to the fact that each Petri net contains unique features and thus has its own application in different areas [7].

The subsequent sections will introduce mathematical notation of the Petri nets used for concurrent control system modeling. They will make use of the base theory with examples, present the most important structural and behavioral properties of Petri nets, and expound on their classes.

2.6 BASIC NOTATIONS AND DEFINITIONS

This section introduces a formal description of Petri nets and their definitions. Notations explain common Petri net properties, especially boundedness and safeness. The definitions correspond to the notation presented in several studies [7, 21, 22, 24, 45, 59, 64].

For better understanding, the collections, sets, and matrices (if applicable) are denoted in uppercase letters, while single variables in lowercase; and vectors are marked by bold style. The multiplication sign is $\cdot$ and $\times$ is a Cartesian product.

Moreover, matrix transposition is represented by ^{T} . A set of natural numbers, including zero, is denoted by $\mathbb{N}$.

Definition 2.9 *A Petri net N* is a 4-tuple:

$$N = (P, T, F, \mathbf{M_0}) , \tag{2.5}$$

where

- $P = \{p_1, p_2, \cdots, p_n\}$ is a set of places,
- $T = \{t_1, t_2, \cdots, t_m\}$ is a set of transitions,
- $F \subseteq (P \times T) \cup (T \times P)$ is a set of arcs,
- $\mathbf{M_0}$ is an initial marking (state) vector.

Graphically, a Petri net is represented as a directed graph whose vertices correspond to places (circles) and edges to transitions (bold line) [24]. Arcs travel from transitions to their output places and from places to their output transitions. Initially, the marked places are pointed by a token (black spot).

Example 2.4 (Petri Net Specification) : Below is an illustration of a Petri net-based specification with a simple example[1]. Fig. (**2.2**) shows a real-life system responsible for controlling a multi-robot. The example presented in a study [45] involves pick-and-place movements to transport or obtain parts by two arms of the robot.

Petri net N_1 consists of $|P_1| = 9$ places and $|T_1| = 6$ transitions. There are nine places denoted by $p_1, \ldots, p_9$, and six transitions marked by $t_1, \ldots, t_6$. A token is put on each place p_1, p_4, p_7, p_8 in Fig. (**2.2**), which means that these places are initially marked: $\mathbf{M_0} = [1\ 0\ 0\ 1\ 0\ 0\ 1\ 1\ 0]$.

Definition 2.10 The place p is an *input place* of the transition $t \in T$, if $(p, t) \in F$.

Definition 2.11 A set of input places of a transition is defined as:

1 Available online: hippo.uz.zgora.pl, *multi robot* benchmark (accessed on November 30, 2024)

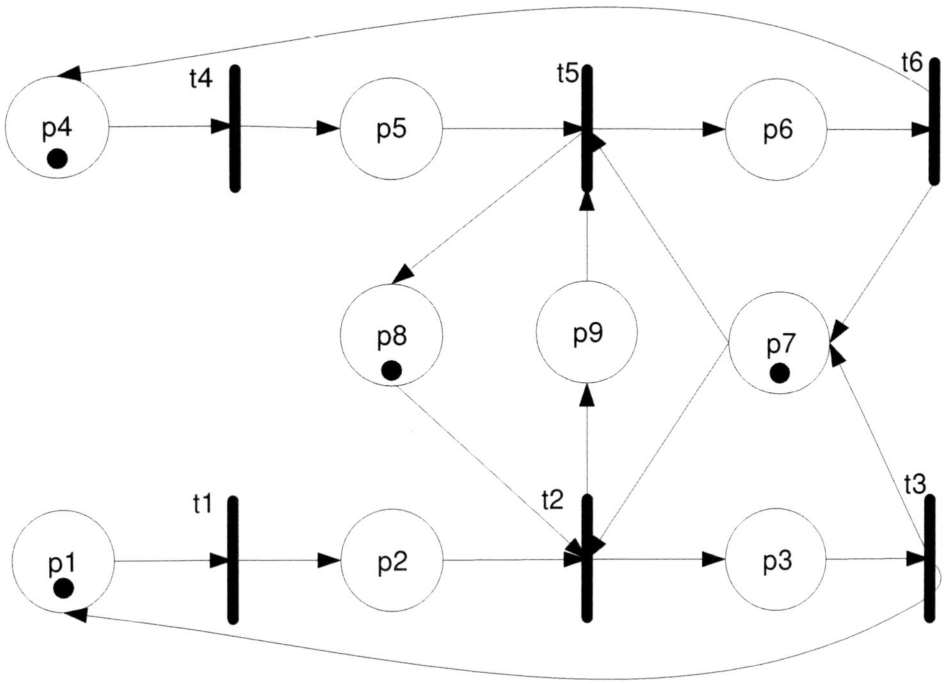

Fig. (2.2). Multi-robot Petri net-based specification N_1.

$$\bullet t = \{p \in P : (p, t) \in F\} . \quad \textbf{(2.6)}$$

Definition 2.12 The place p is an *output place* of the transition $t \in T$, if $(t, p) \in F$.

Definition 2.13 A set of output places of a transition is denoted as:

$$t\bullet = \{p \in P : (t, p) \in F\} . \quad \textbf{(2.7)}$$

Definition 2.14 The transition t is an *input transition* of the place $p \in P$, if $(t, p) \in F$.

Definition 2.15 A set of input transitions of a place is defined as:

$$\bullet p = \{t \in T : (t, p) \in F\}. \quad \textbf{(2.8)}$$

Definition 2.16 The transition t is an *output transition* of the place $p \in P$, if $(p, t) \in F$.

Definition 2.17 A set of output transitions of a place is denoted as:

$$p\bullet = \{t \in T : (p, t) \in F\}. \quad \textbf{(2.9)}$$

Definition 2.18 A transition t is *fired* if each of its input places contains a token. Transition firing removes a token from every input place of t and adds a token to every output place of t.

Definition 2.19 A marking (state) $\mathbf{M_s}$ is *reachable* from marking $\mathbf{M_r}$ if $\mathbf{M_s}$ can be obtained from $\mathbf{M_r}$ by a finite sequence σ of transition firings.

Definition 2.20 A transition t is *enabled* in a marking $\mathbf{M}$ if $\forall p \in \bullet t : M(p) > 0$.

Example 2.5 (Reaching Markings) : Let us look at the multi-robot example once again. Initially, t_1 and t_4 transitions are enabled. After firing transition t_1, the token is moved from its input place p_1 to its output place p_2. Petri net N_1 reaches marking $\mathbf{M_1} = [0\ 1\ 0\ 1\ 0\ 0\ 1\ 1\ 0]$. Note that instead of transition t_1 firing, transition t_4 can also be fired from the initial state $\mathbf{M_0}$ reaching the state $\mathbf{M_2} = [1\ 0\ 0\ 0\ 1\ 0\ 1\ 1\ 0]$. The marking $\mathbf{M_1}$ enables transition t_2, whose execution leads to the next markings.

Definition 2.21 Let R be a *reachability graph* that is a labeled directed graph that represents the state space

$$R = (\mathbb{M}, \Sigma), \quad \textbf{(2.10)}$$

where the vertices of this graph $\mathbb{M}$ are all the reachable markings (states), and the arcs are transitions. An arc $(M_s, M_r) \in \Sigma$ labeled $t_i \in T$ from M_s to M_r exists if firing transition t_i in marking M_s produces marking M_r.

Example 2.6 (Reachability Graph) : Fig. (**2.3**) shows a reachability graph R_1 of a Petri net N_1. In our example of the concurrent control system describing the behavior of a multi-robot, twelve different states can be distinguished in the state space $\mathbb{M}_1$:

$$\begin{array}{l} \quad\quad\;\; p_1\; p_2\; p_3\; p_4\; p_5\; p_6\; p_7\; p_8\; p_9 \\ \mathbf{M_0} = [1\;0\;0\;1\;0\;0\;1\;1\;0] \\ \mathbf{M_1} = [0\;1\;0\;1\;0\;0\;1\;1\;0] \\ \mathbf{M_2} = [1\;0\;0\;0\;1\;0\;1\;1\;0] \\ \mathbf{M_3} = [0\;1\;0\;0\;1\;0\;1\;1\;0] \\ \mathbf{M_4} = [0\;0\;1\;1\;0\;0\;0\;0\;1] \\ \mathbf{M_5} = [0\;0\;1\;0\;1\;0\;0\;0\;1] \\ \mathbf{M_6} = [1\;0\;0\;1\;0\;0\;1\;0\;1] \\ \mathbf{M_7} = [1\;0\;0\;0\;1\;0\;1\;0\;1] \\ \mathbf{M_8} = [0\;1\;0\;1\;0\;0\;1\;0\;1] \\ \mathbf{M_9} = [1\;0\;0\;0\;0\;1\;0\;1\;0] \\ \mathbf{M_{10}} = [0\;1\;0\;0\;1\;0\;1\;0\;1] \\ \mathbf{M_{11}} = [0\;1\;0\;0\;0\;1\;0\;1\;0] \end{array} . \quad \mathbf{(2.11)}$$

Definition 2.22 A Petri net $N = (P, T, F, \mathbf{M_0})$ is claimed to be *k-bounded* or simply *bounded* if

$$\forall p \in P : M(p) \leq k , \quad \mathbf{(2.12)}$$

where $k \in \mathbb{N}$.

It signifies that there is no reachable marking **M** such that any place $p \in P$ contains more than a finite number k of tokens.

Definition 2.23 A Petri net $N = (P, T, F, \mathbf{M_0})$ is said to be *safe* if the number of tokens in each place $p \in P$ does not exceed 1 for any reachable marking **M** that is

$$\forall p \in P : M(p) \leq k = 1 . \quad \mathbf{(2.13)}$$

A safe Petri net is bounded and 1-bounded by the definition.

Definition 2.24 A Petri net $N = (P, T, F, \mathbf{M_0})$ is *L4-live* or just *live* if every transition $t \in T$ can be fired at least once in some firing sequence σ for every reachable marking $\mathbf{M} \in \mathbb{M}$.

Definition 2.25 A Petri net $N = (P, T, F, \mathbf{M_0})$ is *reversible* if for each marking (state) $\mathbf{M} \in \mathbb{M}$, the initial marking $\mathbf{M_0}$ is reachable from that **M**.

Definition 2.26 A Petri net $N = (P, T, F, \mathbf{M_0})$ is known as *well-formed* if it is *safe (1-bounded)*, live and reversible.

Example 2.7 (State Space) : A state space $\mathbb{M}_1$ of a Petri net N_1 shown in Fig. (**2.3**) and described in Equation 2.11 points that for each place $p \in P_1$ in every marking $\mathbf{M} \in \mathbb{M}_1$: $M(p) \leq k = 1$. Therefore, the Petri net N_1 is bounded and safe. Moreover, any transition $t \in T_1$ in any state can be fired in a sequence of firings, thus it is live. In addition, from any reachable marking there is a return path to the initial state $\mathbf{M_0}$ through a sequence of firings, which makes the Petri net reversible. The Petri net N_1 is safe, live, and reversible, which proves that it is well-formed.

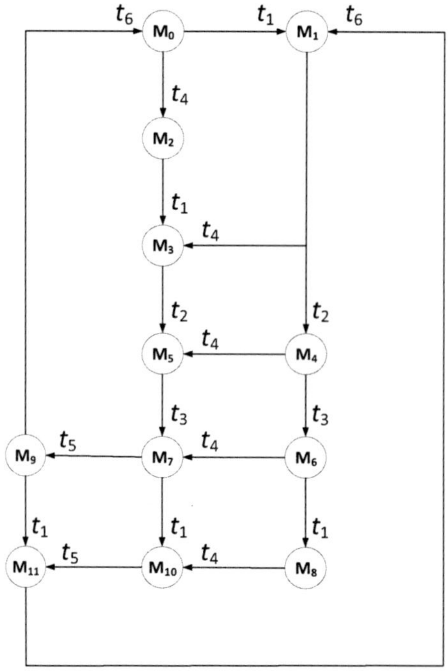

Fig. (2.3). The reachability graph of the Petri net N_1.

Definition 2.27 *An interpreted net* is a *well-formed* Petri net $N = (P, T, F, \mathbf{M_0})$ with additional sets of input and output signals, defined as 6-tuple:

$$N = (P, T, F, \mathbf{M_0}, X, Y) \,, \quad \textbf{(2.14)}$$

where

- $P = \{p_1, p_2, \cdots, p_n\}$ is a set of places,
- $T = \{t_1, t_2, \cdots, t_m\}$ is a set of transitions,
- $F \subseteq (P \times T) \cup (T \times P)$ is a set of arcs,
- $\mathbf{M_0}$ is an initial marking (state) vector,
- $X = \{x_1, x_2, \cdots, x_i\}$ is a set of logic inputs,
- $Y = \{y_1, y_2, \cdots, y_j\}$ is a set of logic outputs.

In contrast to Petri nets, interpreted nets are supported by additional input and output signals, which are used for communication with the environment (real word). Input signals (*e.g.*, received from sensors) are usually related to transitions, while outputs (*e.g.*, actuators) are associated with places [35].

> **Important**

Note that interpreted nets from the definition are *safe (1-bounded)*, live, and reversible Petri nets.

Example 2.8 (Interpreted Net) : An example of an interpreted net is shown in Fig. (**2.4**). This net consists of six places and five transitions. In addition, there are three input signals associated with transitions and six output signals assigned to places. When a place is marked, assigned logical signal is activated (logical high). Moreover, a transition can fire only if an associated signal is high. This finds perfect application in modeling alternate (exclusive-or) paths in a system operation. In this example, t_4 fires only if x_1 is activated. Thus, it changes the marking from p_2 to $p6$, omitting tasks associated with output signals y_3, y_4, y_5.

Definition 2.28 *A deadlock* is a situation in a Petri net $N = (P, T, F, \mathbf{M_0})$ such that from a reachable marking $\mathbf{M_s}$, there is no sequence of firing σ to obtain another reachable marking $\mathbf{M_r}$.

Definition 2.29 A Petri net $N = (P, T, F, \mathbf{M_0})$ is said to be *consistent* if there exist an initial marking (state) $\mathbf{M_0}$ and a firing sequence σ from $\mathbf{M_0}$ back to $\mathbf{M_0}$ such that every transition $t \in T$ occurs at least one in σ.

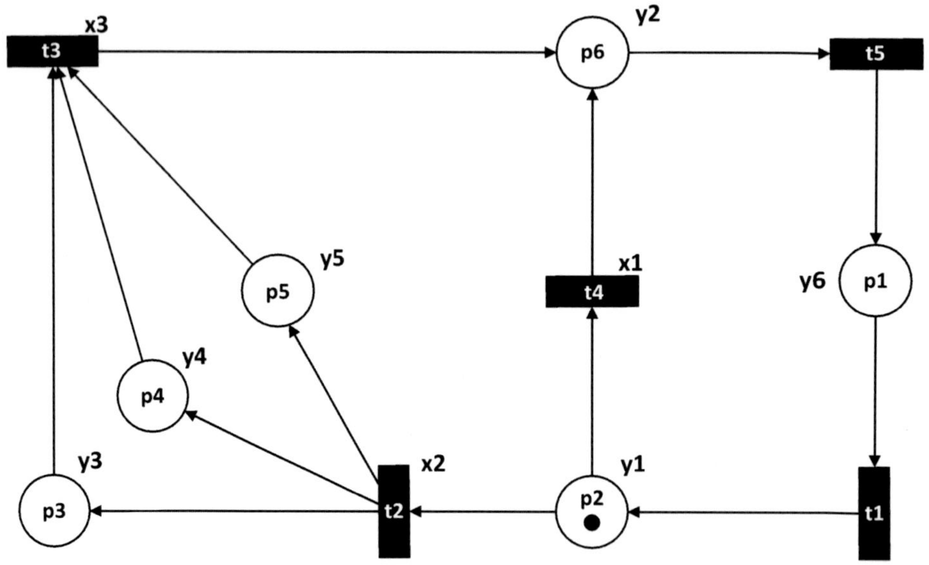

Fig. (2.4). Interpreted net-based specification N_1.

Definition 2.30 A Petri net $N = (P, T, F, \mathbf{M_0})$ is said to be *repetitive* if there exist an initial marking $\mathbf{M_0}$ and a firing sequence σ such that every transition $t \in T$ occurs infinitely often in σ.

Definition 2.31 *An incidence matrix* of a Petri net $N = (P, T, F, \mathbf{M_0})$ with $n = |P|$ places and $m = |T|$ transitions is an $A_{m \times n = [a_{ij}]}$ matrix (where m refers to rows, and n refers to columns) of integers, given by:

$$a_{ij} = \begin{cases} -1, & (p_j, t_i) \in F \\ 1, & (t_i, p_j) \in F \\ 0, & \texttt{otherwise} \end{cases} . \quad \textbf{(2.15)}$$

Example 2.9 (Incidence Matrix) : One of the formal Petri net expressions is an incidence matrix. The incidence matrix of a multi-robot example N_1 is presented in Equation 2.16.

$$A_1 = \begin{array}{c} \begin{matrix} p_1 & p_2 & p_3 & p_4 & p_5 & p_6 & p_7 & p_8 & p_9 \end{matrix} \\ \begin{bmatrix} -1 & 1 & 0 & 0 & 0 & 0 & 0 & 0 & 0 \\ 0 & -1 & 1 & 0 & 0 & 0 & -1 & -1 & 1 \\ 1 & 0 & -1 & 0 & 0 & 0 & 1 & 0 & 0 \\ 0 & 0 & 0 & -1 & 1 & 0 & 0 & 0 & 0 \\ 0 & 0 & 0 & 0 & -1 & 1 & -1 & 1 & -1 \\ 0 & 0 & 0 & 1 & 0 & -1 & 1 & 0 & 0 \end{bmatrix} \begin{matrix} t_1 \\ t_2 \\ t_3 \\ t_4 \\ t_5 \\ t_6 \end{matrix} \end{array} \quad \textbf{(2.16)}$$

The incidence matrix A_1 consists of nine columns representing places. Similarly, transitions are represented by six rows. For example, the cell $a_{23} = 1$ of A_1 determines that p_3 is an output place of transition t_2. Consequently, the cell $a_{57} = -1$ represents the input to the transition t_5 from the place p_7. No connection between a place and a transition is indicated as 0 for example, $a_{25} = 0$. It is easy to see that the graphical structure of the Petri net can be easily represented by this matrix algebraic notation.

Definition 2.32 *A place invariant (p-invariant)* of a Petri net $N = (P, T, F, \mathbf{M_0})$ is a vector

$$\mathbf{x} \geq 0 \text{ such that } A\mathbf{x} = \mathbf{0}, \quad \textbf{(2.17)}$$

where A is the incidence matrix of the net.

Example 2.10 (Place Invariant) : Let us demonstrate an example to clarify what a place invariant stands for. The Petri net N_2 shown in Fig. (**2.5**) represents a dance ballroom.

The Petri net N_2 specifies that there are men and women waiting for their dance in the ballroom. Firing the transition *dance* takes one waiting man and one waiting woman and puts them together on the dance floor, whereas the transition *finish* divides them again. Obviously, the firing transitions (*dance* or *finish*) cannot change the number of persons in the ballroom. Since we can formulate an equation: $1 \cdot M\ (men) + 1 \cdot M\ (women) + 2 \cdot M\ (couples) = \text{const}$. For the initial marking $\mathbf{M_0}$ in the example N_2, we have the following formula: $1 \cdot 1 + 1 \cdot 3 + 2 \cdot 2 = 8$. Note that a nonnegative weight w is assigned to each place $p \in P_2$ so that the weighted token sum remains constant in every reachable state. It is *a place invariant* because the firing of any transition does not change the weighted token sum.

Definition 2.33 *A transition invariant (t-invariant)* of a Petri net $N = (P, T, F, \mathbf{M_0})$ is a vector

$$\mathbf{y} \geq 0 \text{ such that } A^{\mathrm{T}}\mathbf{y} = \mathbf{0}, \quad \textbf{(2.18)}$$

where A^{T} is the transposed incidence matrix A of the net.

Example 2.11 (Transition Invariant) :

The same Petri net example N_2 can illustrate a transition invariant idea. We can formulate a t-invariant when firing transitions in any order result in the same marking as before the series of firings. For example, five firings of transitions *dance* and *finish* return to the same marking: $5 \cdot \ dance \ + 5 \cdot \ finish \ = \text{const}$. Each transition $t \in T$ is assigned a nonnegative weight w such that firing each transition w times does not change the state is *a transition invariant.* Transition invariant corresponds to a cycle in the reachability graph.

Definition 2.34 A Petri net is covered by place invariants if every place $p \in P$ belongs to at least one set of nonzero entries of a place invariant.

Theorem 2.1 *A Petri net* $N = (P, T, F, \mathbf{M}_0)$ *is* bounded *if it is covered by place invariants* [77].

2.7 CLASSES OF PETRI NETS

Wide application of Petri nets, their popularity, and good theoretical background encourage scientists to research their new types and classes. We present classes of

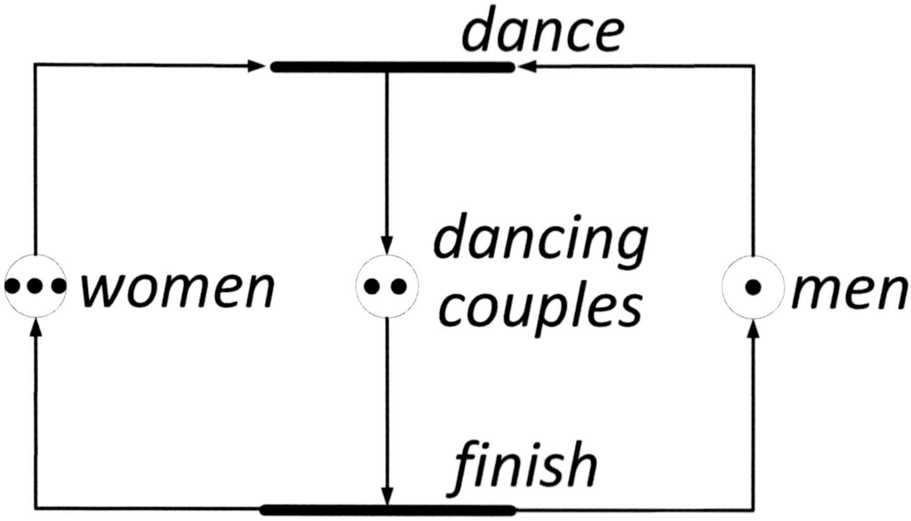

Fig. (2.5). Place invariant idea.

Petri nets in Fig. (**2.6**). The more specific the class, the less complex its analysis, and the more general the class, the better expressiveness it provides. We discuss the problem in detail in Chapter 3, which describes the properties of Petri nets and methods of their analysis.

2.7.1 State Machines

Below are formal definitions related to state machines.

Definition 2.35 *A state machine (SM-net)* is a Petri net $N = (P, T, F, \mathbf{M_0})$ whose transitions $t \in T$ have exactly one input place $p_i \in P$ and one output place $p_o \in P$, *i.e.* $\forall t \in T: |\bullet t| = |t \bullet| = 1$.

Definition 2.36 *A state machine component (SM-component, S-component, SMC)* of a Petri net $N = (P, T, F, \mathbf{M_0})$ is its subnet

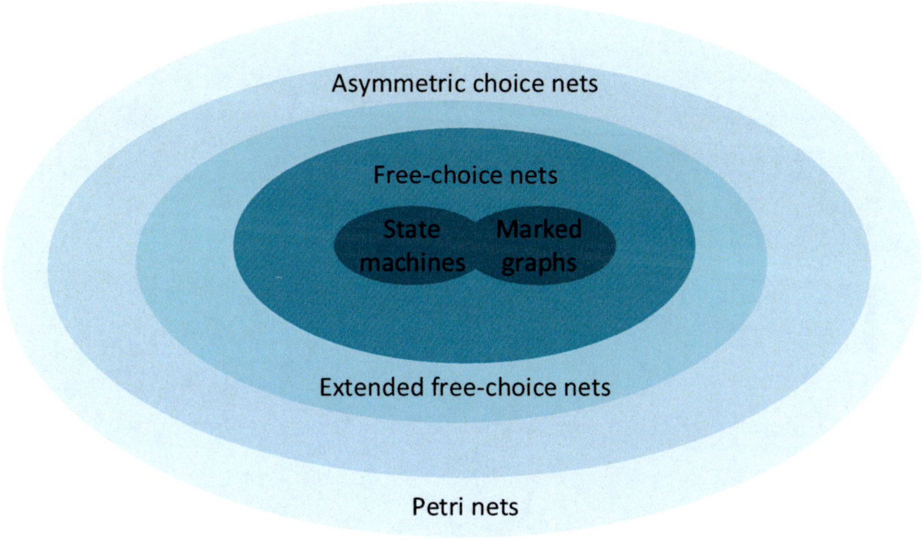

Fig. (2.6). Classes of Petri nets.

$$S = (P', T', F', \mathbf{M}'_0), \quad \textbf{(2.19)}$$

such that S is an SM-net and contains exactly one token in the initial marking.

Definition 2.37 *A state machine cover (SMC-cover)* of a Petri net $N = (P, T, F, \mathbf{M}_0)$ is a set of state machine components

$$\mathbb{S} = \{S_1, \ldots, S_n\}, \quad \textbf{(2.20)}$$

such that each place $p \in P$ is a place of at least one S-component $S \in \mathbb{S}$.

In other words, a state machine is such a Petri net that has just one input and one output place for any transition.

Example 2.12 (State Machine) : See Fig. (**2.7**) as an example of an SM-net. Each transition $t \in \{t_1, t_2, t_3, t_4\}$ has exactly one input and one output place. We can observe that state machines have a sequential structure, and their decision behavior can be modeled. Note that there are two alternate firing sequences in the SM-net presented in Fig. (**2.7**): $\sigma_1 = \{t_1, t_3\}$ or $\sigma_2 = \{t_2, t_4\}$. Since every transition can have just one input and one output place, no concurrency can be specified.

2.7.2 Marked Graphs

Below is a formal description of marked graphs.

Definition 2.38 *A marked graph (MG)* is a Petri net $N = (P, T, F, \mathbf{M}_0)$ such that for all places $p \in P$ holds $|\bullet p| = |p\bullet| = 1$.

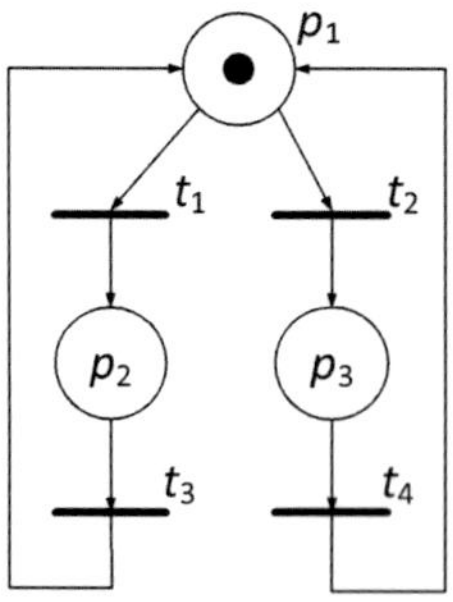

Fig. (2.7). An example of a state machine.

While state machines have one input and one output place of any transitions, marked graphs have exactly one input and one output transition for each place.

Example 2.13 (Marked Graph) : An exemplary marked graph is shown in Fig. (**2.8**) since every place has exactly one input and output transition. Contrary to state machines, marked graphs have a parallel nature and cannot be used to model decision behavior. In the example below, places p_2 and p_3 are concurrent to each other.

2.7.3 Other Classes

Below are remaining definitions and notations that are necessary to explain the idea of free-choice nets, extended free-choice nets, and asymmetric choice nets.

Definition 2.39 *A free-choice net (FCN)* is a Petri net $N = (P, T, F, \mathbf{M_0})$ such that for all places $p \in P$ is true that $|p\bullet| \leq 1 \vee \bullet(p\bullet) = \{p\}$.

Definition 2.40 *An extended free-choice net (EFC)* is a Petri net $N = (P, T, F, \mathbf{M_0})$ such that for all places $p_1, p_2 \in P$, if $p_1\bullet \cap p_2\bullet \neq \emptyset \Rightarrow p_1\bullet = p_2\bullet$.

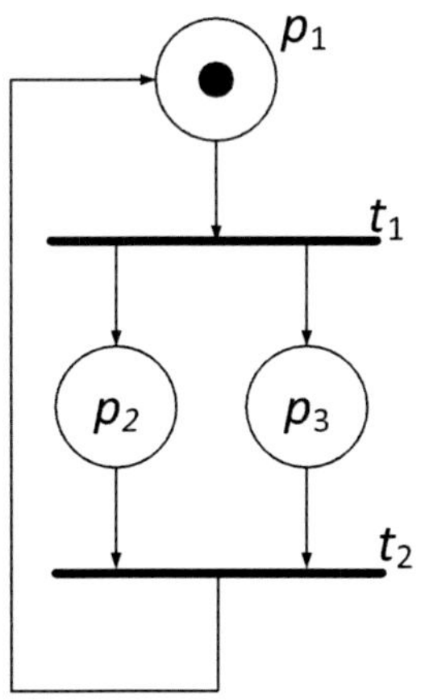

Fig. (2.8) An example of a marked graph.

Definition 2.41 *An asymmetric choice net (ACN)* is a Petri net $N = (P, T, F, \mathbf{M}_0)$ such that for all places $p_1, p_2 \in P$ holds $p_1\bullet \cap p_2\bullet \neq \emptyset \Rightarrow p_1\bullet \supseteq p_2\bullet \vee p_1\bullet \subseteq p_2\bullet$.

2.8 REDUCED ROW ECHELON FORMS

Below are definitions and formal notations explaining the core idea of applying reduced row echelon forms in calculating invariants of Petri nets.

Definition 2.42 A matrix B is said to be in a reduced row echelon form if it satisfies the following conditions [78]:

- if a row of B does not consist entirely of zeros, the first nonzero number in the row is 1 (called a *leading one*),
- any rows that consist entirely of zeros are grouped together at the bottom of B,
- in any two successive non-zero rows of B, the leading one in the lower row occurs farther to the right than the leading one in the higher row,
- each column of B that contains a leading one has zeros everywhere else.

One successful and simple technique for solving linear systems $A\mathbf{x} = \mathbf{0}$ is to reduce the coefficient matrix A of the system to a reduced row echelon form matrix B [78]. This can be accomplished by applying a sequence of elementary row operations like Gauss-Jordan elimination [78]. We know that if A is an $m \times n$ matrix and $\mathbf{x}$ is an n-vector, then the systems $A\mathbf{x} = \mathbf{0}$ and $B\mathbf{x} = \mathbf{0}$ have exactly the same solution set. However, the benefit of such an approach is that the solution set might be read off directly from the matrix B.

There are three elementary row operations to convert any matrix A to its reduced row echelon form matrix B:

- swap two rows,
- multiply a row by any non-zero constant,
- add a scalar multiple of one row to any other row.

The above operations are used in any order to receive a matrix according to Def. 2.42.

Example 2.14 (Linear System) : Consider a simple exemplary linear system given in Equation 2.21.

$$\begin{cases} 5x + 10y - 100 = 0 \\ 40x - 80y = 0 \\ 10x + 20y - 200 = 0 \end{cases} \tag{2.21}$$

And its coefficient matrix in Equation 2.22.

$$A = \begin{bmatrix} 5 & 10 & 100 \\ 40 & -80 & 0 \\ 10 & 20 & 200 \end{bmatrix}. \tag{2.22}$$

Let us now try to solve this linear system in Equation 2.21. Note that one of three equations does not contribute any information to the linear system. The third equation is simply the first one multiplied by two. We can remove the first or third equation without any impact on the solution set.

$$\begin{cases} 5x + 10y - 100 = 0 \\ 40x - 80y = 0 \end{cases} \tag{2.23}$$

The linear system in Equation 2.23 can be solved by any techniques taught in high school. After solving, we know that $x = 10$ and $y = 5$. Now, the reduced row echelon form of matrix A is presented in Equation 2.24.

$$B = \begin{bmatrix} 1 & 0 & 10 \\ 0 & 1 & 5 \\ 0 & 0 & 0 \end{bmatrix} \tag{2.24}$$

For this, we can form a linear system from matrix B:

$$\begin{cases} x + 0 = 10 \\ 0 + y = 5 \\ 0 + 0 = 0 \end{cases} \tag{2.25}$$

The solution can be directly read off: $x = 10$ and $y = 5$. Moreover, a row with all zero entries indicates that one equation in the linear system in Equation 2.21 is redundant.

Subsequently, let us discuss the methodology for invariant computation. As a place invariant is a solution of a linear system $A\mathbf{x} = \mathbf{0}$, the presented approach is adopted for place invariant computation in Petri nets.

Example 2.15 (Reduced Row Echelon Form) : For example, consider an incidence matrix A of Petri net N_3 [2]. The Petri net is shown in Fig. (**2.9**) and its matrix B in a reduced row echelon form accomplished after Gauss-Jordan elimination in Equation 2.27.

$$A = \begin{array}{c} \begin{matrix} p_1 & p_2 \end{matrix} \\ \begin{bmatrix} -1 & 1 \\ -1 & 1 \\ 1 & -1 \end{bmatrix} \end{array}, \tag{2.26}$$

$$B = \begin{array}{c} \begin{matrix} p_1 & p_2 \end{matrix} \\ \begin{bmatrix} 1 & -1 \\ 0 & 0 \\ 0 & 0 \end{bmatrix} \end{array}. \tag{2.27}$$

The solution set can be easily read off:

$$\begin{cases} p_1 + (-p_2) = 0 \\ \qquad 0 = 0 \\ \qquad 0 = 0 \end{cases} \Leftrightarrow p_1 = p_2 \,. \tag{2.28}$$

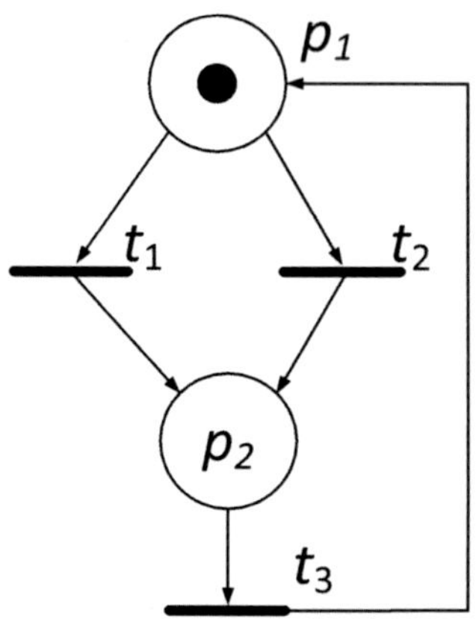

Fig. (2.9). Petri net N_3.

---

[2] Available online: hippo.uz.zgora.pl, *prio_ex* benchmark (accessed on November 30, 2024)

Note that there are just two $\mathbf{x} \geq 0$ integer vectors that do not include other vectors (we call it minimal) and satisfy the system of linear equations $B\mathbf{x} = \mathbf{0}$ and hence $A\mathbf{x} = \mathbf{0}$:

$$\mathbf{x} = [0\ 0]^T \text{ or} \\ \mathbf{x} = [1\ 1]^T . \tag{2.29}$$

We will not consider *a trivial solution* $\mathbf{x} = \mathbf{0}$ as it belongs to a solution set of any homogeneous systems $A\mathbf{x} = \mathbf{0}$, where A is any matrix. Vector $\mathbf{x} = [1\ 1]$ is a place invariant as $A\mathbf{x} = \mathbf{0}$, $\mathbf{x} \geq 0$. We shall see that *some* invariants can be easily read off from the matrix in the reduced row echelon form.

CONCLUSION

We have briefly discussed the base of the Petri net theory supported with simple examples. The most common properties, *i.e.* boundedness and safeness, have been defined. A Petri net-based specification can be used as a modeling language. Owing to parallel nature of nets, the presented Petri net-based methodology is tailored to designing various concurrent control systems. The solution is strongly focused on the model verification aspect. The mathematical structure of Petri nets allows the use of a graph and linear algebra theory for system properties analysis. It is worth noting that other solutions *e.g.*, UML diagrams, do not provide such strict verification tools. It is true that UML state machines benefit from their intuitiveness and common awareness. However, Petri nets provide better model checking and conversion between notations. Moreover, structural conversions enhance Petri nets as a modeling choice, which we will discuss later. On the other hand, exponential time and relatively slow analysis algorithms are the main limitations of the proposed Petri nets. The problem will be discussed in the next chapter.

REFERENCES

[1] T. H. Cormen, C. E. Leiserson, R. L. Rivest, and C. Stein, *Introduction to Algorithms,* 3rd ed. Cambridge, MA: The MIT Press, 2009.

[2] R. Wiśniewski, M. Wiśniewska, and M. Jarnut, "C-exact hypergraphs in concurrency and sequentiality analyses of cyber-physical systems specified by safe petri nets", *IEEE Access*, vol. 7, pp. 13510–13522, 2019.

[3] R. Sedgewick and K. D. Wayne, *Algorithms.* 4th ed., Upper Saddle River, NJ: Addison-Wesley, 2011.

[4] S. Kansal, G. P. Singh, and M. Acharya, "A disconnected 1-safe petri net whose reachability tree is homomorphic to a complete boolean lattice," in *2011 International Conference on Process Automation, Control and Computing*, pp. 1–5, 2011.

[5] Z. Ma, Y. Tong, Z. Li, and A. Giua, "Basis marking representation of petri net reachability spaces and its application to the reachability problem," *IEEE Trans. on Auto. Cont.*, vol. 62, pp. 1078–1093, 2017.

[6] R. J. Wilson, *Introduction to Graph Theory,* 5th ed. Upper Saddle River, NJ: Prentice Hall, 2015.

[7] R. Wiśniewski, G. Bazydło, P. Szcześniak, and M. Wojnakowski, "Petri net-based specification of cyber-physical systems oriented to control direct matrix converters with space vector modulation," *IEEE Access*, vol. 7, pp. 23407–23420, 2019.

[8] R. Wiśniewski, G. Bazydło, and P. Szcześniak, "Low-cost fpga hardware implementation of matrix converter switch control," *IEEE Tran. on Circ. and Sys. II: Exp. Briefs*, vol. 66, pp. 1177–1181, 2019.

[9] G. Bazydło, M. Wojnakowski, and R. Wiśniewski, "The use of UML and Petri net for graphic specification of the reconfigurable logic controllers," *AIP Conference Proceed- ings*, vol. 2040, p. 080004, 2018.

[10] L. Gomes, A. Costa, J. P. Barros, and P. Lima, "From Petri net models to VHDL implementation of digital controllers," in *IECON 2007 - 33rd Annual Conference of the IEEE Industrial Electronics Society*, pp. 94–99, 2007. ISSN: 1553-572X.

[11] W. Liu, P. Wang, Y. Du, M. Zhou, and C. Yan, "Extended logical petri nets-based modeling and analysis of business processes," *IEEE Access*, vol. 5, pp. 16829–16839, 2017.

[12] F. Basile, R. Cordone, and L. Piroddi, "A branch and bound approach for the design of decentralized supervisors in Petri net models," *Automatica*, vol. 52, pp. 322–333, 2015.

[13] Z. Li, "System modeling and control with resource-oriented petri nets by NaiQi Wu and MengChu Zhou," *Inter. J. of Prod. Res.*, Publisher: Taylor & Francis, vol. 49, pp. 6585–6586, 2011.

[14] N. Ran, H. Su, A. Giua, and C. Seatzu, "Codiagnosability analysis of bounded petri nets," *IEEE Tran. on Auto. Con.*, vol. 63, pp. 1192–1199, 2018.

[15] S.-A. Nuño-Sánchez, A. Ramírez-Treviño, and J. Ruiz-León, "Structural sequence detectability in free choice interpreted petri nets," *IEEE Tran. on Auto. Con.*, vol. 61, pp. 198–203, 2016.

[16] M. Szpyrka, J. Biernacki, and A. Biernacka, "Tools and methods for RTCP-nets modeling and verification," *Arch. of Con. Sci.*, Publisher: Committee of Automatic Control and Robotics PAS, no. 3, 2016.

[17] L. Gomes, J. P. Barros, A. Costa, and R. Nunes, "The input-output place-transition petri net class and associated tools," in *2007 5th IEEE International Conference on Industrial Informatics*, vol. 1, pp. 509–514, 2007. ISSN: 2378-363X.

[18] W.-T. Chang, C.-C. Tseng, and W.-K. Chou, "Petri net-based analysis on object assignment in distributed object-oriented systems," *J. of Sys. Arch.*, vol. 44, pp. 955–970, 1998.

[19] L. A. Cortés, P. Eles, and Z. Peng, "Modeling and formal verification of embedded systems based on a Petri net representation," *J. of Sys. Arch.*, vol. 49, pp. 571–598, 2003.

[20] R. Wiśniewski, "Dynamic partial reconfiguration of concurrent control systems specified by petri nets and implemented in xilinx FPGA devices," *IEEE Access*, vol. 6, pp. 32376–32391, 2018.

[21] R. Wiśniewski, M. Wojnakowski, and L. Stefanowicz, "Safety analysis of Petri nets based on the SM-cover computed with the linear algebra technique," *AIP Conference Proceedings*, vol. 2040, p. 080008, 2018.

[22] R. Wiśniewski, A. Karatkevich, L. Stefanowicz, and M. Wojnakowski, "Decomposition of distributed edge systems based on the Petri nets and linear algebra technique," *J. of Sys. Arch.*, vol. 96, pp. 20–31, 2019.

[23] L. Gomes, J. P. Barros, and A. Costa, "Modeling formalisms for embedded system design, embedded systems handbook," in *Embedded Systems Handbook*, Taylor and Francis Group, LLC, 2006.

[24] A. Karatkevich, *Dynamic Analysis of Petri Net-Based Discrete Systems*. Lecture Notes in Control and Information Sciences, Berlin Heidelberg: Springer-Verlag, 2007.

[25] R. David and H. Alla, "Bases of Petri Nets," in *Discrete, Continuous, and Hybrid Petri Nets* R. David and H. Alla, Eds., pp. 1–20, Berlin, Heidelberg: Springer, 2010.

[26] J. Ye, M. Zhou, Z. Li, and A. Al-Ahmari, "Structural decomposition and decentralized control of petri nets," *IEEE Tran. on Sys., Man, and Cyber.: Sys.*, vol. 48, pp. 1360–1369, 2018.

[27] S. S. Peng and M. C. Zhou, "Ladder diagram and Petri-net-based discrete-event control design methods," *IEEE Tran. on Sys., Man, and Cyber., Part C (Applications and Reviews)*, vol. 34, pp. 523–531, 2004.

[28] E. A. Lee, "Cyber Physical Systems: Design Challenges," in *2008 11th IEEE International Symposium on Object and Component-Oriented Real-Time Distributed Comput- ing (ISORC)*, pp. 363–369, 2008. ISSN: 2375-5261.

[29] J. Lee, B. Bagheri, and H.-A. Kao, "A Cyber-Physical Systems architecture for Industry 4.0-based manufacturing systems," *Manufacturing Letters*, vol. 3, pp. 18–23, 2015.

[30] P.-A. Hsiung, T.-W. Kuo, Y.-H. Chang, and C.-H. Huang, "Introduction to the special issue on reconfigurable cyber-physical and embedded system design," *J. of Sys. Arch.*, vol. 62, p. 39, 2016.

[31] Y. Wang, G. Tan, Y. Wang, and Y. Yin, "Perceptual control architecture for cyber physical systems in traffic incident management," *J. of Sys. Arch.*, vol. 58, pp. 398–411, 2012.

[32] M. Abdel-Basset, G. Manogaran, A. Gamal, and F. Smarandache, "A hybrid approach of neutrosophic sets and DEMATEL method for developing supplier selection criteria," *Des Autom Embed Syst*, vol. 22, pp. 257–278, 2018.

[33] R. Wiśniewski, G. Benysek, L. Gomes, D. Kania, T. Simos, and M. Zhou, "IEEE Access Special Section: Cyber-Physical Systems," *IEEE Access*, vol. 7, pp. 157688–157692, 2019.

[34] R. Wiśniewski, G. Bazydło, L. Gomes, A. Costa, and M. Wojnakowski, "Analysis and design automation of cyber-physical system with hippo and IOPT-tools," in *IECON 2019 - 45th Annual Conference of the IEEE Industrial Electronics Society*, vol. 1, pp. 5843–5848, 2019. ISSN: 2577-1647.

[35] R. Wiśniewski, I. Grobelna, and A. Karatkevich, "Determinism in Cyber-Physical Systems Specified by Interpreted Petri Nets," *Sensors*, Number: 19 Publisher: Multidisciplinary Digital Publishing Institute. vol. 20, p. 5565, 2020.

[36] R. Wiśniewski, G. Bazydło, P. Szcześniak, I. Grobelna, and M. Wojnakowski, "Design and Verification of Cyber-Physical Systems Specified by Petri Nets—A Case Study of a Direct Matrix Converter," *Mathematics*, vol. 7, p. 812, 2019.

[37] Z. Li, M. Zhou, and N. Wu, "A survey and comparison of Petri net-based deadlock prevention policies for flexible manufacturing systems," *IEEE Tran. on Sys., Man, and Cyber., Part C (Applications and Reviews)*, vol. 38, pp. 173–188, 2008.

[38] A. Ramírez-Trevino, E. Ruiz-Beltran, I. Rivera-Rangel, and E. Lopez-Mellado, "Online fault diagnosis of discrete event systems. A petri net-based approach," *IEEE Tran. on Auto. Sci. and Eng.*, vol. 4, pp. 31–39, 2007.

[39] W. M. P. van der Aalst, K. M. van Hee, A. H. M. ter Hofstede, N. Sidorova, H. M. W. Verbeek, M. Voorhoeve, and M. T. Wynn, "Soundness of workflow nets: Classification, decidability, and analysis," *Form Asp Comp*, vol. 23, pp. 333–363, 2011.

[40] N. Ran, S. Wang, and W. Wu, "Event feedback supervision for a class of petri nets with unobservable transitions," *IEEE Access*, vol. 6, pp. 6920–6926, 2018.

[41] F. Yang, N. Wu, Y. Qiao, and R. Su, "Polynomial approach to optimal one-wafer cyclic scheduling of treelike hybrid multi-cluster tools *via* Petri nets," *IEEE/CAA J. of Auto. Sinica*, vol. 5, pp. 270–280, 2018.

[42] S. M. Vahidipour, M. Esnaashari, A. Rezvanian, and M. R. Meybodi, "GAPN-LA: A framework for solving graph problems using Petri nets and learning automata," *Eng. Applic. of Arti. Intel.*, vol. 77, pp. 255–267, 2019.

[43] I. Grobelna, "Model checking of reconfigurable FPGA modules specified by Petri nets," *J. of Sys. Archi.*, vol. 89, pp. 1–9, 2018.

[44] I. Grobelna, R. Wiśniewski, M. Grobelny, and M. Wiśniewski, "Design and verification of real-life processes with application of petri nets," *IEEE Transactions on Systems, Man, and Cybernetics: Systems*, vol. 47, pp. 2856–2869, 2017.

[45] M. Wojnakowski and R. Wiśniewski, "Verification of the boundedness property in a Petri net-based specification of the control part of cyber-physical systems," *in Technological Innovation for Applied AI Systems*, L. M. Camarinha-Matos, P. Ferreira, and G. Brito, Eds., IFIP Adv. Inf. Commun. Technol., Cham: Springer Int. Publ., pp. 83–91, 2021.

[46] C. Girault and R. Valk, *Petri Nets for Systems Engineering: A Guide to Modeling, Verification, and Applications*. Berlin Heidelberg: Springer-Verlag, 2003.

[47] X. He, Z. Dong, H. Yin, and Y. Fu, "A framework for developing cyber-physical systems," *Int. J. Soft. Eng. Knowl. Eng.*, vol. 27, pp. 1361–1386, 2017.

[48] R. Cavada, A. Cimatti, M. Dorigatti, A. Griggio, A. Mariotti, A. Micheli, S. Mover, M. Roveri, and S. Tonetta, "The nuXmv symbolic model checker," in *Computer Aided Verification* A. Biere and R. Bloem, Eds., Lecture Notes in Computer Science, Cham: Springer International Publishing, pp. 334–342, 2014.

[49] M. Szpyrka, "Analysis of VME-Bus communication protocol - RTCP-net approach," *Real-Time Syst*, vol. 35, pp. 91–108, 2007.

[50] M. Malhotra and K. Trivedi, "Dependability modeling using Petri-nets," *IEEE Trans. Reliab.*, vol. 44, pp. 428–440, 1995.

[51] X. Zhang and S. Yao, "Fuzzy stochastic Petri nets and analysis of the reliability of multi-state systems," *IET Software*, vol. 9, no. 3, pp. 83–93, 2015.

[52] G. Y. Liu, Z. W. Li, K. Barkaoui, and A. M. Al-Ahmari, "Robustness of deadlock control for a class of Petri nets with unreliable resources," *Infor. Sci.*, vol. 235, pp. 259–279, 2013.

[53] M. Abdel-Basset, M. Gunasekaran, M. Mohamed, and F. Smarandache, "A novel method for solving the fully neutrosophic linear programming problems," *Neural Comput & Applic*, vol. 31, pp. 1595–1605, 2019.

[54] M. Abdel-Basset, G. Manogaran, A. E. Fakhry, and I. El-Henawy, "2-Levels of clustering strategy to detect and locate copy-move forgery in digital images," *Multimed Tools Appl*, vol. 79, pp. 5419–5437, 2020.

[55] Y. Sun, B. McMillin, X. Liu, and D. Cape, "Verifying noninterference in a cyber-physical system the advanced electric power grid," in *7th International Conference on Quality Software. (QSIC 2007)*, pp. 363–369, 2007. ISSN: 2332-662X.

[56] M. Alcaraz-Mejia, E. Lopez-Mellado, and A. Ramírez-Trevino, "Redundancy based controller reconfiguration for fault recovery of manufacturing systems," in *2007 IEEE International Conference on Automation Science and Engineering*, pp. 128–133, 2007. ISSN: 2161-8089.

[57] M. Abdel-Basset, G. Manogaran, L. Abdel-Fatah, and S. Mirjalili, "An improved nature inspired meta-heuristic algorithm for 1-D bin packing problems," *Pers Ubiquit Comput*, vol. 22, pp. 1117–1132, 2018.

[58] M. Abdel-Basset, G. Manogaran, H. Rashad, and A. N. H. Zaied, "A comprehensive review of quadratic assignment problem: variants, hybrids and applications," *J Ambient Intell Human Comput*, 2018.

[59] R. Wiśniewski, *Prototyping of Concurrent Control Systems Implemented in FPGA Devices*. Advances in Industrial Control, Springer International Publishing, 2017.

[60] F. Pereira, F. Moutinho, J. P. Barros, A. Costa, and L. Gomes, *Executable models for Embedded Controllers Development — A Cloud Based Development Framework*. 2015.

[61] C. A. Petri, *Kommunikation mit Automaten*, Ph.D. thesis, Mathematisches Institut der Universität Bonn, Bonn, OCLC: 258511501, 1962.

[62] G. Callou, P. Maciel, D. Tutsch, J. Araújo, J. Ferreira, and R. Souza, "Petri nets - manufacturing and computer science", *A Petri Net-Based Approach to the Quantification of Data Center Dependability*, IntechOpen, 2012.

[63] A. Gholami and B. S. Bigham, "A learned soccer goalkeeper Petri net model," in *2017 Artificial Intelligence and Robotics (IRANOPEN)*, pp. 102–108, 2017.

[64] T. Murata, "Petri nets: Properties, analysis and applications," *Proc. of the IEEE*, vol. 77, pp. 541–580, 1989.

[65] K. Barkaoui and M. Minoux, "A polynomial-time graph algorithm to decide liveness of some basic classes of bounded Petri nets," in *Application and Theory of Petri Nets 1992* (K. Jensen, ed.), Lecture Notes in Computer Science, Berlin, Heidelberg, Springer, pp. 62–75, 1992.

[66] J. Patalas-Maliszewska, R. Wiśniewski, M. Topczak, and M. Wojnakowski, "Design optimization of the Petri net-based production process supported by additive manufac- turing technologies," *Bull. Pol. Acad. Sci. Tech. Sci.*, vol. 70, no. 2, p. e140693, 2022.

[67] R. David and H. Alla, *Discrete, Continuous, and Hybrid Petri Nets*. 2nd Ed., Berlin Heidelberg: Springer-Verlag, 2010.

[68] L. Holloway and B. Krogh, "Synthesis of feedback control logic for a class of controlled Petri nets," *IEEE Trans. Autom. Control*, vol. 35, pp. 514–523, 1990.

[69] H. Li, "Petri net as a formalism to assist process improvement in the construction industry," *Autom. Constr.*, vol. 7, pp. 349–356, 1998.

[70] F. Moutinho and L. Gomes, "Asynchronous-channels within petri net-based GALS distributed embedded systems modeling," *IEEE Trans. Ind. Informat.*, vol. 10, pp. 2024–2033, 2014.

[71] M. Szpyrka, P. Matyasik, R. Mrówka, and L. Kotulski, "Formal description of alvis language with α 0 system layer," *Fundam. Inform.,* IOS Press, vol. 129, pp. 161–176, 2014.

[72] W. M. P. Van Der Aalst, "The application of petri nets to workflow management," *J. Circuits Syst. Comput.*, Publisher: World Scientific Publishing Co, vol. 08, pp. 21–66, 1998.

[73] M. Diaz, "Applying petri net based models in the design of systems," in *Concurrency and Nets: Advances in Petri Nets* K. Voss, H. J. Genrich, and G. Rozenberg, Eds., pp. 23–67, Berlin, Heidelberg: Springer, 1987.

[74] L. Ferrarini, "Computer aided design of logic controllers with petri nets," in *Petri Nets in Flexible and Agile Automation*, M. Zhou, Ed., The Springer International Series in Engineering and Computer Science, pp. 71–92, Boston, MA: Springer US, 1995.

[75] R. David and H. Alla, "Autonomous continuous and hybrid petri nets," in *Discrete, Continuous, and Hybrid Petri Nets* R. David and H. Alla, Eds., pp. 117–158, Berlin, Heidelberg: Springer, 2010.

[76] R. David and H. Alla, "Timed continuous petri nets," in *Discrete, Continuous, and Hybrid Petri Nets,* R. David and H. Alla, Eds., pp. 159–229, Berlin, Heidelberg: Springer, 2010.

[77] W. Reisig, "Nets consisting of places and transistions," in *Petri Nets: An Introduction* , W. Reisig, Ed., EATCS Monographs on Theoretical Computer Science, Berlin, Heidelberg: Springer, pp. 62–76, 1985.

[78] A. T. Yuster and T. Yuster, "The reduced row echelon form of a matrix is unique: A simple proof," *Math. Mag.*, 1984.

CHAPTER 3

Boundedness and Safeness

Abstract: This chapter provides an in-depth overview of related work. Boundedness and safeness are at the core of various methodologies using the theory of Petri nets. The problems of analyzing the properties of Petri nets, such as boundedness and safeness, were undertaken more than 30 years ago and are still the subject of research. Boundedness is an important feature of various systems, such as manufacturing systems. A bounded Petri net determines a finite number of reachable states.

Keywords: Boundedness, Coverability graphs, Petri net invariants, Reachability graphs, Safeness.

3.1 INTRODUCTION

Let us imagine a control system specified by a Petri net that is in charge of an elevator. If the Petri net were unbounded, the elevator could behave in an indefinite way and enter states not foreseen by the designer. For example, the elevator could stop between floors or go higher than is allowed and thus cause mechanical failure. Moreover, a safe Petri net may be used to model systems aimed at hardware implementation in microcontrollers or FPGAs — logical control (token/ no token) [1, 2]. In the elevator example, such logical behavior could make the elevator go up or could disable going up.

The most popular analysis techniques [3-6] that allow one to check the boundedness and safeness of a Petri net can be generally divided into two main groups [7]: based on reachability graphs and structural analyzes using linear algebra methodology. The first approach is based on the exploration of reachability graphs.

Upon constructing a state space, each subsequent possible reachable state is placed on the graph. In general, the number of reachable states of a system can grow exponentially (a state explosion problem). It translates into exponential run-time of algorithms analyzing properties on reachability graphs. Therefore, for some Petri nets consisting of up to several dozen places or transitions, the methods may not produce a result in the assumed time, *i.e.*, they are not efficient in such cases. The second approach is based on the linear algebra technique [2, 8-12]. In this approach, all minimal invariants of a Petri net are computed (minimal means that they are not contained in other invariants). Place invariants are used by linear algebraic procedures to check the boundedness and safeness. However, the number of all minimal invariants can also grow exponentially. Therefore, both verification paths have serious limitations because the number of reachable states and place invariants in the graph can be exponential [1, 13, 14]. This makes the computational complexity of algorithms investigating these properties exponential, which means that a solution may not be found within the assumed reasonable time.

Below, problems of reachability graph building, computing invariants, and analyses, which attempt to handle these problems by means of popular methods of boundedness and safeness, are presented. In addition, computer-aided tools used in concurrent control specification based on Petri nets are mentioned.

3.2 COVERABILITY GRAPHS

Finite graphs can be prepared in which every reachable marking is explicitly represented by a node [15]. Such graphs are known as coverability graphs [13]. Petri net properties can be directly obtained from the exploration of such graphs. Such analyses can guarantee that a Petri net is bounded or unbounded, safe or unsafe, or whether it contains deadlocks. However, this approach can be inefficient as the number of reachable states can be exponential. This means that the operation of the algorithm may not be completed in the assumed time. Net reduction, *i.e.*, simplification of a Petri net with preservation of its boundedness, safeness, and other behavior, can be one of the practical solutions to this problem [1]. However, such reduced graphs can still be too large to be computed [1, 16]. The size of the reductions depends on the characteristics of a Petri net. Some Petri nets can be significantly reduced to make it possible to analyze them efficiently. However, other Petri nets may reduce marginally or not at all, which makes it impossible to verify them.

3.2.1 Overview

As Lipton proved in a study [17], the building of a reachability graph requires an exponential space and a decidable marking of its reachability problem [18]. Although space hardness has been shown for Petri nets in general [17], for some classes of Petri nets, *e.g.*, acyclic nets, the state reachability problem can be resolved by means of an integer linear programming problem. Moreover, it can be solved in polynomial time in state machines and marked graphs. The reachability problem is part of the Petri net theory. Many properties, including boundedness and safeness, can be resolved by a system reachability analysis.

In the following sections, the most interesting properties will be described to proceed with the construction techniques of the reachability graphs. Binary decision diagrams (BDD) can be used for some properties using the symbolic traverse technique. However, since the BDD data structure is based on place invariants, reconstructing the behavior of a Petri net is not straightforward [19].

Valmari's set of methods [20] is perfectly applicable to Petri nets with identical concurrent structures. For instance, various identical workflows are triggered simultaneously, and they run in parallel. However, the requirement that all workflows have an identical structure, which is a very restrictive presumption, limits the applicability of this approach. A subnet that reduces into a single transition with preservation properties such as boundedness is presented in some articles [21, 22]. In a study [23], the author creates a dependency graph for removing some substructures without changing the property of reachability. However, the limitation of these approaches is that they are highly based on some particular substructures of Petri nets.

Recently, a state compression approach [24, 25] has been proposed by Cabaisnoet *et al.* The benefit of this methodology is that only a part of the reachability space (called basis markings) is enumerated. This technique can be effectively used to solve controllability problems [26] or opacity problems [27, 28] in Petri nets. In addition, numerous different methods with new classes of Petri nets are developed to handle the marking reachability problem. For example, siphon analysis is applied to determine if deadlock markings exist in S^3PR nets [29-31].

In a paper [19], a practically efficient algorithm based on a reachability graph solves the marking reachability problem in Petri nets. This method has a wide relevance, since it does not depend on particular substructures of the Petri nets. Reachability sets can be precisely characterized by this technique.

There are tools available that implement the generation of state space to determine the properties of a Petri net. There are a great number of computer-aided design software programs that support the specification of Petri nets. Up to 100 such programs can be identified. Many of them are solutions presented in the 1990s, *e.g.*, PAPETRI, Xpetri, PROD, INA, SPNP, and others. The websites of these programs do not respond and probably their development was completed many years ago. Another group is programs whose last version was created in the years 2000–2010, among which we can distinguish programs, such as Yasper [32] without the option of analyzing Petri nets, JSARP [33] with analysis of only boundedness and liveness on reachability graphs, JARP with a state space generator, but without information on properties. The most popular and cited in recent publications with a wide range of verification are IOPT-Tools [2, 34-42] and PIPE [43-45]. Both CAD programs perform a check of properties such as boundedness and safeness based on reachability graphs. Due to the state explosion problem, some Petri nets with several or more places and transitions cannot be checked in the assumed time.

3.2.2 Construction

The construction of a reachability graph that contains all the reachable markings of a Petri net is shown in Algorithm 3.1. This fundamental method proposed by Murata [13] is claimed to be a reference reachability graph determining algorithm. The paper [13] also introduces the symbol ω as an indicator of a place that accumulates tokens infinitely in order to maintain a finite reachability graph (coverability graph) for unbounded Petri nets.

Building a coverability graph (finite reachability graph) begins with initial marking as a current state. In turn, it is checked whether it is possible to reach a new state from this marking, *i.e.*, if there are enabled transitions. Then, each subsequent enabled transition is fired, and a new state is generated. The new state becomes a current marking and further new states are searched for according to the same procedure. After each new marking, if such a state $\mathbf{M_r}$ is obtained on the path from the initial marking $\mathbf{M_0}$ to the current marking $\mathbf{M_s}$ that $\forall p \in P, M_s(p) \geq M_r(p)$, then a symbol ω is placed for $p \in P, M_s(p) > M_r(p)$, indicating that for those places, tokens will accumulate in an infinite way.

Example 3.1 (Construction of a Coverability Graph) : The reachability graph shown in Fig. (**2.3**) of the Petri net N_1 was obtained after performing the following steps described in Algorithm 3.1. The first state is the initial marking M_0. We then check whether it is possible to gain a new state by seeking the enabled transitions from the

```
   Data: A Petri net N = (P, T, F, M0)
   Result: The coverability graph R = (𝕄, Σ)
1  M ← an initial marking M0;
2  while exists a new marking that can be reachable from M do
3      foreach enabled transition t at M do
4          Ms ← a marking after t fired from M;
5          if Ms ∉ 𝕄 then
6              if exists Mr on the path from M0 that ∀p ∈ P, Ms(p) ≥ Mr(p) then
7                  foreach p ∈ P, Ms(p) > Mr(p) do
8                      Ms(p) ← ω;
9                  end
10             end
11             𝕄 ← 𝕄 ∪ {Ms};
12             Σ ← Σ ∪ {(M, Ms)};
13             M ← Ms;
14         end
15     end
16 end
```

Algorithm 3.1: Construction of a coverability graph.

current state $\mathbf{M_0}$. In the case of the $\mathbf{M_0}$ state, two transitions are enabled to fire: t_1 and t_4. Therefore, each enabled transition is fired in turn, and a new state is determined. In our example, after firing t_1, we get the marking M_1, and after transition t_4, the marking M_2. For each established state, we check the unbounded place condition (line 6 of Algorithm 3.1). For the M_1 and M_2 states, all places are bounded —— the condition does not apply. In the same way, subsequent states are determined until no transition is enabled or a new state cannot be resolved (all of them have been placed in the graph).

3.3 INVARIANTS

Computation of invariants is one of the most popular techniques for verifying and analysis of Petri nets [1, 13, 46, 47]. Place invariants can be used to analyze important properties of Petri nets, such as boundedness and safeness [1, 2, 8, 13, 46-49]. It should be emphasized that checking the properties of Petri nets by obtaining invariants is usually more efficient than exploring the reachability graph, especially for larger nets, *i.e.*, with a dozen places or transitions and more.

The analysis of boundedness requires computation of a place-invariant cover of the Petri net [8, 50]. Reisig [50] proves that the place-invariant Petri nets covered are *bounded* and their transition invariant is repetitive. Repetitiveness can be used for the investigation of significant properties, such as liveness or absence of deadlocks. Also, according to a study [13], each t-invariant of a Petri net is p-invariant of a dual net. Notice that a dual net has a transposed incidence matrix. A Petri net is said to be covered by place or transition invariants if its every place or transition belongs to at least one set of corresponding nonzero entries in a p- or t-invariant, respectively.

The most trivial approach to find minimal place invariants $\mathbf{x} \geq 0$ that $A\mathbf{x} = \mathbf{0}$ is testing all $\mathbf{x}$ permutations in domain $x\ (i) \in \{0,\ 1\}$, where $i\ \in \{1..n\}$. Similarly, minimal transition invariants can be found when solving equation: $A^{T}\mathbf{y} = \mathbf{0} \Leftrightarrow \mathbf{y}A = \mathbf{0}$. By minimal invariants we mean those that contain no other invariants. Each invariant can be multiplied by an integer to get another invariant. Thus, whenever we search for invariants, we refer to the minimal ones, *i.e.*, those that are not contained in others. Algorithm 3.2 shows a trivial procedure to obtain place invariants for any Petri net.

Data: A Petri net $N = (P, T, F, \mathbf{M_0})$
$A \leftarrow$ an incidence matrix of N;
$C \leftarrow$ `Find All Binary Permutations` $(|P|)$;
foreach *vector* $\mathbf{x}$ *of matrix* C **do**
    if $A \times \mathbf{x} = \mathbf{0}$ **then**
        print "vector $\mathbf{x}$ is the p-invariant of net N";
    end
end

Algorithm 3.2: Place invariants calculating a trivial procedure.

This algorithm is obviously non-efficient because we need $2^{|P|}$ equality check operations. Assuming one matrix multiplication takes 10^{-4} seconds [1], for a middle-size Petri net (*e.g.*, $|P| = 30$), the whole procedure lasts approx. 30 hours. The application of this method for a 43-place Petri net requires as much time as the age of the author of this book.

[1] Octave software on AMD Phenom II X4 965 processor (3.4 GHz x 4) and 12 GB of RAM

In general, the linear algebra approach to obtain place invariants can be treated as a *Boolean satisfiability problem*. The problem is represented as $p_1 + p_2 + \cdot \cdot \cdot + p_n = 0$, where $p_i, i \in \{1..n\}$ is a Boolean variable, and thus is *NP-complete* [51]. In another study [51], Cook proved that any problem in *NP class* can be reduced in polynomial time by *a deterministic Turing machine*. An important corollary of the described theorem is that *Boolean satisfiability* is equivalent to the *P versus NP problem*, which is widely considered the most important unsolved problem in theoretical computer science [52].

The first invariant computing method is proposed in a research work [53]. All invariants are produced one by one. The algorithm can be stopped at any time of its execution. While polynomial amount of memory is used, the computation time in the worst case is exponential.

The Fourier-Morzkin method [46] is another well-known proposition for computing all invariants. It has the following critical drawbacks. Even if invariants exist, none of them may be computed because of an overflow caused by candidate vectors for invariants whose number may become exponential or many non-minimal invariants may be included too. There are several ideas to improve the Fourier-Morzkin method, which have been suggested in some papers [46, 54-56]. As for this approach, only one or a few minimal invariants can be found instead of all. It may shorten computation time and save memory.

A relatively new algorithm GMST is presented in a study [57]. It makes it possible to compute at least one invariant in short computation time under condition that any invariant exists in an inputted Petri net. Schmidt [58] worked out a sketch of an algorithm to compute invariants of algebraic nets. His method has a number of disadvantages, including intolerable time complexity and computational problems [58].

A Toudic method algorithm [59] features means of matrix transformations for obtaining non-negative integer solutions (invariants). However, the complexity of this solution is exponential, which restricts the analysis of real-life models [60]. Zaitsev implements this algorithm in a study [61], and based on the decomposition of Petri nets, he improves it in another study [60] to reach exponential acceleration with respect to the number of places. The same author summarizes in 2006 [62] that when investigating properties of Petri nets, all well-known methods display exponential computational complexity.

Finally, the most popular and commonly used algorithm for computation of invariants was initially introduced in a research work [46]. The Martinez-Silva method is widely known and described in many papers [8, 48, 49]. It also constitutes a basis for other methods of computing invariants or improving this reference algorithm. The procedure is presented step by step in Algorithm 3.3.

Data: A Petri net $N = (P, T, F, \mathbf{M_0})$
Result: Rows of D that correspond to the found p-invariants
$D \leftarrow$ an identity matrix of size $|P|$;
$A \leftarrow$ an incidence matrix of N;
$A^{\mathrm{T}} \leftarrow$ `Transpose Matrix` (A);
$Q \leftarrow$ union matrix of D and A^{T};
for each *column j of matrix* A^T **do**
    find row pairs that their sum is equal to 0 and add it to Q;
    delete rows that intersection with j-th column is not equal to 0;
    reduce redundant rows of Q;
end

Algorithm 3.3: Martinez-Silva's algorithm for calculating place invariants.

Algorithm 3.3 is based on algebra (matrix operations). Firstly, a unit matrix Q is formed. This matrix conjugates two matrices: an identity matrix D of places number size of a Petri net and a transposed incidence matrix A^{T} of the net. Note that matrix D will hold the obtained place invariants after the execution of the algorithm. Subsequently, for each column of the transposed incidence matrix, the algorithm searches for row pairs that annul the adequate column. These rows are appended to matrix Q. Then, non-minimal invariants, being supersets, are removed from Q. Finally, rows of A^{T} that contain all zeroes are related to the proper place invariant in matrix D.

However, exponential computational complexity is the main limitation in the analysis of invariants by this method and its derivatives, since it makes it possible to obtain all the minimal invariants of a Petri net [1]. Moreover, place invariants are presented after the algorithm terminates its operation.

3.4 PROPERTY OF BOUNDEDNESS

Apart from safeness, this property is the most common subject matter in surveys on Petri nets [63]. A bounded Petri net has a finite number of tokens for each place in any reachable state [3, 64]. In turn, the number of tokens is nondeterministic in Petri nets that are unbounded.

Example 3.2 (Unbounded Petri Net) : Fig. (**3.1**) shows the Petri net N_4 [2] given as an example of an unbounded net. We can see that in the firing loop sequence $\sigma = \{t_1, t_2, t_1, t_2, t_1, t_2, \cdot \cdot \cdot \}$, the new tokens are produced in place p_3 ad infinitum.

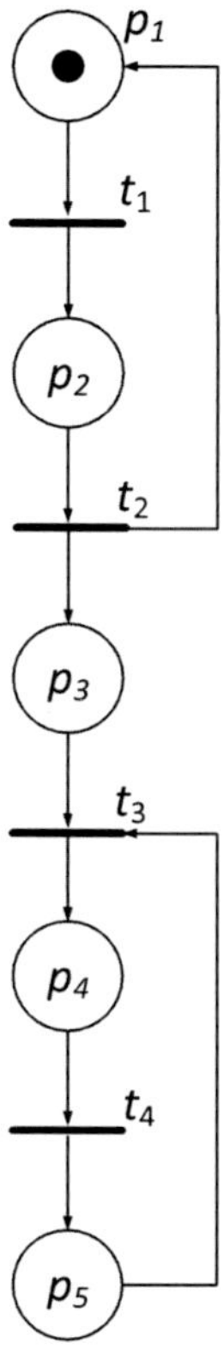

Fig. (3.1) An example of an unbounded Petri net.

[2] Available online: hippo.uz.zgora.pl, *consumerReachability* benchmark (accessed on November 30, 2024)

Table 3.1 Minimal and maximal bounds of the Petri net N_5.

-	p_1 (population)	p_2 (axe)	p_3 (woodcutter)	p_4 (iron)	p_5 (coal)
min bounds	2	0	0	0	4
max bounds	10	8	8	3	7

If a Petri net is unbounded, the graph representation grows infinitely large [13]. Unbounded Petri nets are a wide topic that requires different analytic tools, so it is out of the scope in our discussion. In a study [65], authors address this subject and propose complex reachability trees to analyze such nets. Generally, such unbounded nets have little practical application in cases studied within this book.

Example 3.3 (Bounded Petri Net): Fig. (**3.2**) shows a bounded Petri net N_5. The idea of this example is based on the famous video game series — The Settlers [3]. In the Petri net with $|P_5| = 5$ places and $|T_5| = 3$ transitions, we can see ten tokens in the place p_1 (*population*) representing ten free settlers. There are also five available *axes* (p_2), three *iron* (p_4) bars, and seven units of *coal* (p_5). During the game (when changing the states), if a player needs to *recruit* (fired t_2) a *woodcutter* (p_3), we have to provide one free settler from the population (p_1) with one axe (p_2). When the woodcutter (p_3) is *retired* (t_1), he comes back to the free population (p_1) and returns his axe (p_2). At the same time, a new axe can be made by *Toolsmith's Works* (t_3) from iron (p_4) and coal (p_5).

Table **3.1** presents minimal and maximal numbers of objects (axes, iron, coal, *etc.*) in any time of the game. Note that the number of objects is always (in any reachable state) *bounded*. In other words, the object neither disappears, nor is it preternaturally multiplied.

Algorithm 3.4 presents boundedness analysis by means of exploration of reachability graphs. The function *Construct a Coverability Graph* is described in Algorithm 3.1. It should be remembered that for some Petri nets this approach to the construction of graphs cannot be completed in reasonable time because of computational complexity.
Boundedness analysis with the application of Martinez-Silva invariants computation method is shown in Algorithm 3.5. Obtaining place invariants is proceeded by

[3] More information available online: https://en.wikipedia.org/wiki/The_Settlers (accessed on November 30, 2024)

the function *Calculate Place Invariants* detailed in Algorithm 3.3. The whole procedure essentially checks the coverage of a Petri net by the obtained place invariants. If a Petri net is covered, then it is bounded.

Below, application and usefulness of bounded Petri nets in recent research and in various proposed solutions are presented. Kaid *et al.* proposes to apply Petri nets to the scheduling and deadlock analysis in reconfigurable manufacturing systems [3]. The algorithm uses the most important properties of Petri nets, including boundedness. The author's technique permits dynamic modification of the structure of the manufacturing system without affecting its behavioral properties (boundedness, liveness, reversibility).

The solution presented in a study [66] proposes deadlock recovery policy in flexible manufacturing systems. Generally, the proposed algorithm performs building reachability graphs of Petri nets to meet specification requirements, such as reachability or liveness. The task requires verification of boundedness for the designed concurrent control system.

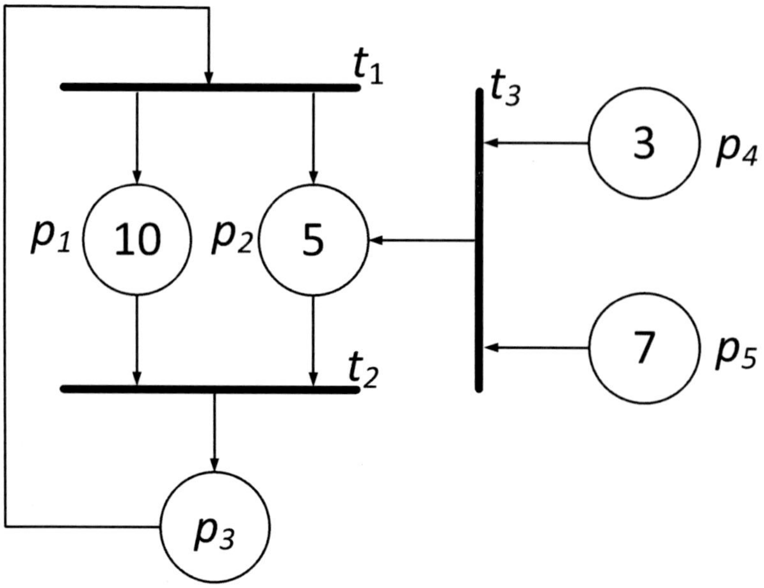

Fig. (3.2). An example of a bounded Petri net.

Data: A Petri net $N = (P, T, F, \mathbf{M_0})$
Result: The net N is bounded or unbounded
$R = (\mathbb{M}, \Sigma) \leftarrow$ `Construct a Coverability Graph` (N);
for each *state* **M** *at* $\mathbb{M}$ **do**
    for each *place* p *at* **M** **do**
        if $M(p) = \omega$ **then**
            is_bounded $\leftarrow$ false;
        end
    end
end
if *is_bounded* **then**
    print "net N is bounded";
else
    print "net N is unbounded";
end

Algorithm 3.4: Boundedness analysis based on a coverability graph.

Data: A Petri net $N = (P, T, F, \mathbf{M_0})$
Result: The net N is bounded or uncovered
$I \leftarrow$ `Calculate Place Invariants` (N);
for each *column p of matrix I* **do**
    if $p = 0$ **then**
        is_covered $\leftarrow$ false;
    end
end
if *is_covered* **then**
    print "net N is bounded";
else
    print "net N is uncovered by place invariants";
end

Algorithm 3.5: Boundedness analysis based on place invariants cover.

A system specified by a bounded Petri net is necessary for the diagnosability analysis technique described in a study [67]. The paper also involves verification of reachability graphs of Petri nets. In order to reduce the computational complexity due to the state explosion problem, a special class of Petri nets known as "modified verifier nets" is introduced. An automaton "verifier" makes it possible to detect malfunctions of the modeled system.

Bounded Petri nets are applied in another study [68] for a synthesis algorithm. The system is composed of smaller sub-components — transition systems. A symbolic representation of states is involved to generate such components to be synthesized. The authors experimentally verified their algorithm and implemented it as a tool. It was shown that application of bounded Petri nets resulted in significant reduction in the synthesis.

Decomposition methods of persistent Petri nets are introduced in a study [69]. The paper states that bounded and reversible nets can be decomposed into smaller, transition-disjoint cycles. The authors propose several methods based on reachability graphs and linear algebra techniques. Boundedness of Petri nets is a crucial property and is required in such analyses.

The property of boundedness finds its application in an algorithm analyzing so-called "non-blockingness" of a specified system. This property is especially often utilized in the supervisory control theory of discrete-event systems [4]. The proposed method is meant to reduce analytical complexity by avoiding the space explosion problem. This is accomplished with integer linear algebra as opposed to constructing a traditional reachability graph.

Concluding the above, it can be observed that the property of boundedness is essential in design, analysis, verification, and also decomposition of Petri net-based concurrent control systems. Most practices require a bounded Petri net as their input data for further processing. Moreover, bounded Petri nets find their application when some kind of memory is necessary, contrary to safe nets which fit into *e.g.*, logical control.

3.5 PROPERTY OF SAFENESS

Safeness can be treated as stronger boundedness. Safe Petri nets are also called 1-bounded, because the maximal number of tokens for each place is one in any reachable state. Places in safe Petri nets display binary behavior: there is one token in a place or there is no token in a place for any reachable marking. Because of this feature, safe Petri nets find their application in interpreted nets that have additional sets of logical input/ output signals.

Multiple publications all over the world feature the research on safeness [1]. The basic theory regarding safety analysis of Petri nets is introduced in several studies [1, 70, 71]. Zaitsev [62] provides a necessary and sufficient condition for the property of safeness

$$\exists \mathbf{x} > 0, A\mathbf{x} = \mathbf{0}\,. \tag{3.1}$$

Like boundedness, safeness can be tested on coverability graphs, as presented in Algorithm 3.6. The advantages and limitations of the solution are the same as in the case of analysis of boundedness.

```
Data: A Petri net N = (P, T, F, M0)
Result: The net N is safe or unsafe
R = (𝕄, Σ) ← Construct a Coverability Graph (N);
foreach state M at 𝕄 do
    foreach place p at M do
        if M(p) > 1 then
            is_safe ← false;
        end
    end
end
if is_safe then
    print "net N is safe";
else
    print "net N is unsafe";
end
```

Algorithm 3.6: Safeness analysis based on a coverability graph.

Moreover, the Martinez-Silva method of calculating invariants can also be adapted to the verification of safeness. Instead of covering place invariants, Algorithm 3.7 checks that the resulting invariants form correct state machine components (ref. Def. 2.36) and whether a given Petri net is covered by these S-components.

Some more advanced algorithms checking safeness are proposed and shown in some studies [1, 72-75]. However, they are applicable for particular classes of Petri nets [1, 7, 72-75]. For instance, in marked graphs, the net that is live is safe if each place $p \in P_{MG}$ belongs to a cycle, which possesses exactly one place that is marked under $\mathbf{M}_{MG}$. For this class of Petri net, analysis problem is time polynomial. A parallel algorithm implemented in the HiPS tool [76] can be used for efficient (according to authors) safeness analysis of free-choice net class only.

```
Data: A Petri net N = (P, T, F, M0)
Result: The net N is safe or uncovered
I ← Calculate Place Invariants (N);
foreach row x of matrix I do
    if x is not correct SMC then
        I ← I − {x};
    end
end
foreach column p of matrix I do
    if p = 0 then
        is_SMCs_covered ← false;
    end
end
if is_SMCs_covered then
    print "net N is safe";
else
    print "net N is uncovered by SMCs";
end
```

Algorithm 3.7: Safeness analysis based on SMCs cover.

Silva *et al.* [77] showed a noteworthy linear programming approach to the safeness checking problem. This property is satisfied if the following equation is bounded and its optimal value is less than or equal to $k = 1$.

$$\min\{\mathbf{x}^{\mathrm{T}} M_0 | \mathbf{x}^{\mathrm{T}} A \leq 0 ; \mathbf{x} \geq \mathbf{e_p}\}, \qquad (3.2)$$

where $\mathbf{e_p}$ is a vector with $e_p[p_i] = 0$ for all $p_i \neq p$ and $e_p[p] = 1$. Thus, linear programming problems can be solved by well-known polynomial time algorithms [77]. Nevertheless, the property of safeness cannot be concluded if the optimal value of the linear programming equation is greater than k [77].

Several design solutions require a safe Petri net as an input to their methods, for instance, those presented in several studies [5, 68, 78-80]. The authors of a research work [81] proposed the modeling of a discrete event system based on a safe Petri net. With the theory of supervisory control, a specification that requires avoiding a set of forbidden markings is typically considered. The proposed algorithm designs a maximally permissive deadlock avoidance controller for a safe Petri net system. Furthermore, apart from safeness, other control problems are analyzed in the paper.

Based on the theory of general regions, Carmona *et al.* [78] present an algorithm for bounded or safe Petri net synthesis. An efficient synthesis methodology for concurrent control systems is introduced. Similar presumptions are treated in another study [82], *i.e.*, a safe Petri net is obligatory as an input to construct a controller. Next, maximal permissive controllers are obtained by computation of invariants. The work introduces a systematic algorithm to depreciate the number of linear constraints corresponding to the forbidden states. This is performed by means of non-reachable states and by building constraints using a systematic scheme.

In a paper [83], a given Petri net needs to be safe, while transitions of the net can have labels with symbols from a considered alphabet. Moreover, safe nets are also appropriate for verification aspects. The authors claim that every system of finite states can be expressed as a safe-labeled Petri net, and as a proof, a novel synthesis algorithm is proposed.

Safe Petri nets for a natural and widespread model of concurrent control systems are considered in a study [84]. It proposes a notion of "pullback" for a Petri net, and according to the author, pullbacks can be mainly useful to design two concurrent control systems. These controllers share some resources and are synchronized by means of common events. The paper grants a simple construction for pullbacks of safe Petri nets.

A paper [79] offers efficient concurrency and sequentiality analysis solutions committed to the control part of a cyber-physical system specified by a safe Petri net. The introduced idea is based on the hypergraph theory and combines computation of subsequent exact transversals in a c-exact hypergraph in a polynomial time. The proposed method is supported by adequate theorems, algorithms, and proofs.

Profits of specification with the application of safe Petri nets are shown in a study [80]. The authors discuss analytical methods of computational complexity for three Petri net classes: acyclic nets, conflict-free nets, and free-choice nets. All the analyzed Petri nets are required to be safe. Safe Petri nets are also subject of research in another study [85], where a safe Petri net is converted into a colored Petri net and again into an uncolored net to prove that all the nets in this construction have the same partially ordered multisets.

The authors in a paper [5] submitted a model verification tool called *AdamMC*. This tool is devoted to safe Petri nets that acquire a Petri net-based specification as its input in PNML (Petri net markup language) format, extended with LTL (linear-time temporal logic) forms connected to places and transitions of the Petri net. The proposed algorithms reduce computational complexity by including sequential and parallel optimization processes.

Coverability analysis of a system specified by Petri nets is proposed in a study [86]. The algorithm is based on the utilization of a satisfiability modulo theories solver to produce an inductive invariant. The presented model checking analysis uses integer arithmetic. Furthermore, the author discusses similar methodologies using experimental evaluation of methods.

Summarizing the above discussion, like boundedness, safeness is a principle and a relevant property in modeling and decomposition of Petri net-based concurrent control systems. Many of the presented techniques require safe Petri nets as input data in algorithms and tools for further verification and implementation.

CONCLUSION

The most common problems in the analysis of Petri nets, such as the reachability problem and computation of invariants, along with the properties of boundedness and safeness, have been studied in this chapter. A broad literature review, regarding the application of bounded or safe Petri nets, has also been presented, together with a survey of algorithms useful for verifying the properties. The chapter also provides descriptions of known algorithms for constructing reachability graphs or computing all minimal invariants. These algorithms will serve as reference methods in the experimental research presented in Chapter 5, because the Petri net-based approach can be used as a modeling language. The aim of the discussion is to design various concurrent control systems.

REFERENCES

[1] R. Wiśniewski, *Prototyping of Concurrent Control Systems Implemented in FPGA Devices, in Advances in Industrial Control*, Springer International Publishing, 2017.

[2] R. Wiśniewski, G. Bazydło, L. Gomes, A. Costa, and M. Wojnakowski, "Analysis and Design Automation of Cyber-Physical System with Hippo and IOPT-Tools," in *IECON 2019 - 45th Annual Conference of the IEEE Industrial Electronics Society*, vol. 1, pp. 5843–5848, 2019. ISSN: 2577-1647.

[3] H. Kaid, A. Al-Ahmari, Z. Li, and R. Davidrajuh, "Automatic supervisory controller for deadlock control in reconfigurable manufacturing systems with dynamic changes," *Applied Sciences*, Multidisciplinary Digital Publishing Institute. vol. 10, p. 5270, 2020.

[4] C. Gu, Z. Ma, Z. Li, and A. Giua, "Verification of nonblockingness in bounded petri nets with a semi-structural approach," in *2019 IEEE 58th Conference on Decision and Control (CDC)*, pp. 6718–6723, 2019. ISSN: 2576-2370.

[5] B. Finkbeiner, M. Gieseking, J. Hecking-Harbusch, and E.-R. Olderog, "AdamMC: A model checker for petri nets with transits against flow-LTL," in *Computer Aided Verification,* S. K. Lahiri and C. Wang, Eds., Lecture Notes in Computer Science, Cham: Springer International Publishing, pp. 64–76, 2020.

[6] I. Grobelna, R. Wiśniewski, and M. Wojnakowski, "Specification of cyber-physical systems with the application of interpreted nets," in *IECON 2019 - 45th Annual Conference of the IEEE Industrial Electronics Society*, vol. 1, pp. 5887–5891, 2019. ISSN: 2577-1647.

[7] K. Barkaoui and M. Minoux, "A polynomial-time graph algorithm to decide liveness of some basic classes of bounded Petri nets," in *Application and Theory of Petri Nets 1992,* K. Jensen, Ed., Lecture Notes in Computer Science, Berlin, Heidelberg: Springer, pp. 62–75, 1992.

[8] M. Wojnakowski and R. Wiśniewski, "Verification of the boundedness property in a petri net-based specification of the control part of cyber-physical systems," in *Technological Innovation for Applied AI Systems,* L. M. Camarinha-Matos, P. Ferreira, and G. Brito, Eds., IFIP Advances in Information and Communication Technology, Cham: Springer International Publishing, pp. 83–91, 2021.

[9] R. Wiśniewski, G. Bazydło, P. Szcześniak, and M. Wojnakowski, "Petri net-based specification of cyber-physical systems oriented to control direct matrix converters with space vector modulation," *IEEE Access*, vol. 7, pp. 23407–23420, 2019.

[10] M. Wojnakowski, M. Popławski, R. Wiśniewski, and G. Bazydło, "Safeness analysis of petri net-based cyber-physical systems based on the linear algebra and parallel reductions," in *AIP Conference Proceedings*, Heraklion, Greece, 2021.

[11] M. Wojnakowski, R. Wiśniewski, G. Bazydło, and M. Popławski, "Analysis of safeness in a Petri net-based specification of the control part of cyber-physical systems," *AMCS*, vol. 31, pp. 647–657, 2021.

[12] M. Wojnakowski, M. Popławski, R. Wiśniewski, and G. Bazydło, "Hippo-CPS: verification of boundedness, safeness and liveness of petri net-based cyber-physical systems," in *Technological Innovation for Digitalization and Virtualization,* L. M. Camarinha-Matos, Ed., IFIP Advances in Information and Communication Technology, Cham: Springer International Publishing, pp. 74–82, 2022.

[13] T. Murata, "Petri nets: Properties, analysis and applications," *Proceedings of the IEEE*, vol. 77, pp. 541–580, 1989.

[14] A. Karatkevich, *Dynamic Analysis of Petri Net-Based Discrete Systems*. Lecture Notes in Control and Information Sciences, Berlin Heidelberg: Springer-Verlag, 2007.

[15] W. Reisig, *Petri Nets: An Introduction*. Monographs in Theoretical Computer Science. An EATCS Series, Berlin Heidelberg: Springer-Verlag, 1985.

[16] J. Esparza and M. Silva, "Top-down synthesis of live and bounded free choice nets," in *Advances in Petri Nets 1991,* G. Rozenberg, Ed., Lecture Notes in Computer Science, Berlin, Heidelberg: Springer, pp. 118–139, 1991.

[17] R. J. Lipton, "The reachability problem requires exponential space," in *Technical Report 62*, Yale University, 1976.

[18] E. Mayr, "Persistence of vector replacement systems is decidable," *Acta Informatica*, vol. 15, pp. 309–318, 1981.

[19] Z. Ma, Y. Tong, Z. Li, and A. Giua, "Basis marking representation of petri net reach- ability spaces and its application to the reachability problem," *IEEE Trans. Autom. Control.*, vol. 62, pp. 1078–1093, 2017.

[20] A. Valmari, "Stubborn sets for reduced state space generation," in *Proceedings of the 10th International Conference on Applications and Theory of Petri Nets: Advances in Petri Nets 1990*, Berlin, Heidelberg: Springer-Verlag, pp. 491–515, 1991.

[21] F. Dicesare, G. Harhalakis, J.-M. Proth, M. Silva-Suarez, and F. Vernadat, *Practice of Petri Nets in Manufacturing*. Springer Netherlands, 1993.

[22] I. Suzuki and T. Murata, "A method for stepwise refinement and abstraction of Petri nets," *J. Comput. Syst. Sci.*, vol. 27, pp. 51–76, 1983.

[23] P. Küngas, "Petri net reachability checking is polynomial with optimal abstraction hierarchies," in *Abstraction, Reformulation and Approximation,* J.-D. Zucker and L. Saitta, Eds., Lecture Notes in Computer Science, Berlin, Heidelberg: Springer, pp. 149–164, 2005.

[24] A. Giua and C. Seatzu, "Fault detection for discrete event systems using Petri nets with unobservable transitions," in *Proceedings of the 44th IEEE Conference on Decision and Control*, pp. 6323–6328, 2005. ISSN: 0191-2216.

[25] M. P. Cabasino, A. Giua, M. Pocci, and C. Seatzu, "Discrete event diagnosis using labeled Petri nets. An application to manufacturing systems," *Control Eng. Pract.*, vol. 19, pp. 989–1001, 2011.

[26] Z. Ma, Z. Li, and A. Giua, "A method to verify the controllability of language specifi- cations in Petri nets based on basis marking analysis," in *54th IEEE Conference on Decision and Control (CDC)*, pp. 1675–1681, 2015.

[27] Y. Tong, Z. Li, C. Seatzu, and A. Giua, "Verification of initial-state opacity in Petri nets," in *2015 54th IEEE Conference on Decision and Control (CDC)*, pp. 344–349, 2015.

[28] Y. Tong, Z. Li, C. Seatzu, and A. Giua, "Verification of current-state opacity using Petri nets," in *2015 American Control Conference (ACC)*, pp. 1935–1940, 2015. ISSN: 2378-5861.

[29] J. Ezpeleta, J. Colom, and J. Martinez, "A Petri net based deadlock prevention policy for flexible manufacturing systems," *IEEE Trans. Rob. Autom.*, vol. 11, pp. 173–184, 1995.

[30] Z. Li and M. Zhou, "Elementary siphons of Petri nets and their application to deadlock prevention in flexible manufacturing systems," *IEEE Trans. on Sys., Man, and Cybern. - Part A: Sys. and Huma.*, vol. 34, pp. 38–51, 2004.

[31] Z. Li and M. Zhou, "Control of Elementary and Dependent Siphons in Petri Nets and Their Application," *IEEE Trans. on Sys., Man, and Cybern. - Part A: Sys. and Huma.,*, vol. 38, pp. 133–148, 2008.

[32] K. van Hee, O. Oanea, R. Post, L. Somers, and J. van der Werf, "Yasper: a tool for workflow modeling and analysis," in *6th International Conference on Application of Concurrency to System Design (ACSD'06)*, pp. 279–282, 2006. ISSN: 1550-4808.

[33] F. G. Oliviera Lino, *Analisador e Simulador de Redes de Petri.* Bachelor Thesis, University of Rio de Janeiro, Rio de Janeiro, 2007.

[34] L. Gomes, F. Moutinho, and F. Pereira, "IOPT-tools — A Web based tool framework for embedded systems controller development using Petri nets," in *2013 23rd International Conference on Field programmable Logic and Applications*, pp. 1–1, 2013. ISSN: 1946-1488.

[35] F. Pereira, F. Moutinho, J. P. Barros, A. Costa, and L. Gomes, *Executable models for Embedded Controllers Development — A Cloud Based Development Framework*, 2015.

[36] L. Gomes, A. Costa, J. P. Barros, and P. Lima, "From Petri net models to VHDL implementation of digital controllers," in *IECON 2007 - 33rd Annual Conference of the IEEE Industrial Electronics Society*, pp. 94–99, 2007. ISSN: 1553-572X.

[37] L. Gomes, J. P. Barros, A. Costa, and R. Nunes, "The input-output place-transition petri net class and associated tools," in *2007 5th IEEE International Conference on Industrial Informatics*, vol. 1, pp. 509–514, 2007. ISSN: 2378-363X.

[38] F. Pereira, F. Moutinho, and L. Gomes, "IOPT tools user manual - version 1.1," 2014.

[39] R. Campos-Rebelo, F. Pereira, F. Moutinho, and L. Gomes, "From IOPT Petri nets to C: An automatic code generator tool," in *2011 9th IEEE International Conference on Industrial Informatics*, pp. 390–395, 2011. ISSN: 2378-363X.

[40] L. Gomes and J. P. Barros, "Refining IOPT petri nets class for embedded system controller modeling," in *IECON 2018 - 44th Annual Conference of the IEEE Industrial Electronics Society*, pp. 4720–4725, 2018. ISSN: 2577-1647.

[41] F. Pereira, F. Moutinho, L. Gomes, and R. Campos-Rebelo, "IOPT Petri net state space generation algorithm with maximal-step execution semantics," in *2011 9th IEEE International Conference on Industrial Informatics*, pp. 789–795, 2011. ISSN: 2378-363X.

[42] F. Pereira, F. Moutinho, and L. Gomes, "IOPT-tools — Towards cloud design automation of digital controllers with Petri nets," in *2014 International Conference on Mechatronics and Control (ICMC)*, pp. 2414–2419, 2014.

[43] P. Bonet and C. Lladó, "PIPE v 2 .5: a petri net tool for performance modeling," 2007.

[44] N. J. Dingle, W. J. Knottenbelt, and T. Suto, "PIPE2: a tool for the performance evaluation of generalised stochastic Petri Nets," *SIGMETRICS Perform. Eval. Rev.*, vol. 36, pp. 34– 39, 2009.

[45] W. Knottenbelt, "PIPE v2. 5: A Petri net tool for performance modelling," in *Proceedings of the 23rd Latin American Conference on Informatics (CLEI 2007)*, Osijek, Croatia, Faculty of Law, Josip Juraj Strossmayer University in Osijek, vol. 12, 2007.

[46] J. Martínez and M. Silva, "A simple and fast algorithm to obtain all invariants of a generalised petri net," in *Application and Theory of Petri Nets,* C. Girault and W. Reisig, Eds., Informatik-Fachberichte, Berlin, Heidelberg: Springer, pp. 301–310, 1982.

[47] B. Li, M. Khlif-Bouassida, and A. Toguye′ni, "On–the–fly diagnosability analysis of bounded and unbounded labeled petri nets using verifier nets," *AMCS, Int. J. Appl. Math. Comput. Sci.*. vol. 28, pp. 269–281, 2018.

[48] R. Wiśniewski, A. Karatkevich, L. Stefanowicz, and M. Wojnakowski, "Decomposition of distributed edge systems based on the Petri nets and linear algebra technique," *J. Syst. Archit.*, vol. 96, pp. 20–31, 2019.

[49] R. Wiśniewski, M. Wojnakowski, and L. Stefanowicz, "Safety analysis of Petri nets based on the SM-cover computed with the linear algebra technique," *AIP Conference Proceedings*, vol. 2040, p. 080008, 2018.

[50] W. Reisig, "Nets consisting of places and transistions," in *Petri Nets: An Introduction,* W. Reisig, Ed., EATCS Monographs on Theoretical Computer Science, Berlin, Heidelberg: Springer, pp. 62–76, 1985.

[51] S. A. Cook, "The complexity of theorem-proving procedures," in *Proceedings of the 3rd annual ACM symposium on Theory of computing*, STOC '71, New York, NY, USA: Association for Computing Machinery, pp. 151–158, 1971.

[52] R. M. Karp, "Reducibility among Combinatorial Problems," in *Complexity of Computer Computations: Proceedings of a symposium on the Complexity of Computer Computations,* R. E. Miller, J. W. Thatcher, and J. D. Bohlinger, Eds., The IBM Research Symposia Series, Boston, MA: Springer US, pp. 85–103, 1972.

[53] T. Tanida, T. Watanabe, and K. Onaga, "A polynomial-time algorithm for finding a semi- generator of Petri net invariants," in *1991 IEEE International Symposium on Circuits and Systems (ISCAS)*, pp. 2838–2841 vol. 5, 1991.

[54] J. M. Colom and M. Silva, "Convex geometry and semiflows in P/T nets. A comparative study of algorithms for computation of minimal p-semiflows," in *Advances in Petri Nets 1990,* G. Rozenberg, Ed., Lecture Notes in Computer Science, Berlin, Heidelberg: Springer, pp. 79–112, 1991.

[55] K. Takano, S. Taoka, M. Yamauchi, and T. Watanabe, "Experimental evaluation of two algorithms for computing petri net invariants," *IEICE TRANSACTIONS on Fundamentals of Electronics, Communications and Computer Sciences*, The Institute of Electronics, Information and Communication Engineers. vol. E84-A, pp. 2871–2880, 2001.

[56] K. Takano, "Efficient computation of nonnegative integer-invariants of Petri nets," in *International Conference on Fundamentals of Electronics. Communications and Com- puter Sciences (ICFS 2002)*, pp. S5/19–24, IEICE, 2002.

[57] A. Taguchi, S. Taoka, and T. Watanabe, "An algorithm GMST for extracting minimal siphon-traps and its application to efficient computation of Petri net invariants," in *Proceedings of the 2003 International Symposium on Circuits and Systems, 2003. ISCAS '03.*, vol. 3, pp. 3–3, 2003. ISSN: null.

[58] K. Schmidt, "On the computation of place invariants for algebraic petri nets," in *Structures in Concurrency Theory,* J. Desel, Ed., Workshops in Computing, (London), pp. 310–325, Springer, 1995.

[59] J. M. Toudic, "Linear algebra algorithms for the structural-analysis of petri nets,"*Revue Technique Thomson-CSF*, vol. 14, no. 1, pp. 137–155, 1982.

[60] D. Zaitsev, "Decomposition-based calculation of Petri net invariants," in *Workshop on Token based computing*, (Bologna, Italy), 2004.

[61] D. Zaitsev, "Formal grounding of Toudic method," (Eichstaett, Germany), 2003.

[62] D. A. Zaitsev, "Compositional analysis of Petri nets," *Cybern Syst Anal*, vol. 42, pp. 126– 136, 2006.

[63] K. P. Valavanis, "On the hierarchical modeling analysis and simulation of flexible manufacturing systems with extended Petri nets," *IEEE Trans. on Sys., Man, and Cybern.*, vol. 20, pp. 94–110, 1990.

[64] J. Patalas-Maliszewska, R. Wiśniewski, M. Topczak, and M. Wojnakowski, "Design optimization of the Petri net-based production process supported by additive manufac- turing technologies," *Bull. Pol. Acad. Sci. Tech. Sci.*, vol. 70, no. 2, p. e140693, 2022.

[65] F. Lu, Q. Zeng, M. Zhou, Y. Bao, and H. Duan, "Complex reachability trees and their application to deadlock detection for unbounded petri nets," *IEEE Trans. Syst. Man Cybern. Syst.*, vol. 49, pp. 1164–1174, 2019.

[66] Y. Chen, Z. Li, A. Al-Ahmari, N. Wu, and T. Qu, "Deadlock recovery for flexible manufacturing systems modeled with Petri nets," *Inf. Sci.*, vol. 381, pp. 290–303, 2017.

[67] N. Ran, J. Hao, Z. He, and C. Seatzu, "Diagnosability analysis of bounded Petri nets," in *2018 IEEE 23rd International Conference on Emerging Technologies and Factory Automation (ETFA)*, vol. 1, pp. 1145–1148, 2018. ISSN: 1946-0740.

[68] E. Badouel, L. Bernardinello, and P. Darondeau, "Polynomial algorithms for the synthesis of bounded nets," in *TAPSOFT '95: Theory and Practice of Software Development,* P. D. Mosses, M. Nielsen, and M. I. Schwartzbach, Eds., Lecture Notes in Computer Science, Berlin, Heidelberg: Springer, pp. 364–378, 1995.

[69] E. Best and P. Darondeau, "Decomposition theorems for bounded persistent Petri nets," in *Applications and Theory of Petri Nets,* K. M. van Hee and R. Valk, Eds., Lecture Notes in Computer Science, Berlin, Heidelberg: Springer, pp. 33–51, 2008.

[70] F. Commoner, A. W. Holt, S. Even, and A. Pnueli, "Marked directed graphs," *J. Comput. Syst. Sci.*, vol. 5, pp. 511–523, 1971.

[71] J. B. Dennis, "Bibliography," in *Record of the Project MAC conference on concurrent systems and parallel computation*, pp. 183–199, New York, NY, USA: Association for Computing Machinery, 1970.

[72] E. Best, "Structure theory of petri nets: The free choice hiatus," in *Petri Nets: Central Models and Their Properties,* W. Brauer, W. Reisig, and G. Rozenberg, Eds., Lecture Notes in Computer Science, Berlin, Heidelberg: Springer, pp. 168–205, 1987.

[73] E. Best and P. S. Thiagarajan, "Some classes of live and safe Petri nets," in *Concurrency and nets: advances in Petri nets*, Berlin, Heidelberg: Springer-Verlag, pp. 71–94, 1987.

[74] P. Kemper, "O(|P||T|)-Algorithm to Compute a Cover of S-components in EFC-nets," 1995.

[75] M. H. T. Hack, "Analysis of production schemata by Petri nets," *tech. rep., Massachusetts Inst. Tech., Cambridge Project MAC,* Section: Technical Reports, 1972.

[76] K. Ide and K. Wasaki, "Efficient algorithm for liveness/safeness analysis of free-choice nets in a Petri net tool: HiPS," in *2012 6th International Conference on New Trends in Information Science, Service Science and Data Mining (ISSDM2012)*, pp. 179–184, 2012.

[77] M. Silva, J. M. Colom, J. Campos, and G. C, "Linear algebraic techniques for the analysis of Petri nets," in *In: Recent Advances in Mathematical Theory of Systems, Control, Networks, and Signal Processing II*, Mita Press, pp. 35–42, 1992.

[78] J. Carmona, J. Cortadella, M. Kishinevsky, A. Kondratyev, L. Lavagno, and A. Yakovlev, "A symbolic algorithm for the synthesis of bounded Petri nets," in *Applications and Theory of Petri Nets,* K. M. van Hee and R. Valk, Eds., Lecture Notes in Computer Science, Berlin, Heidelberg: Springer, pp. 92–111, 2008.

[79] R. Wiśniewska, M. Wiśniewska, and M. Jarnut, "C-exact hypergraphs in concurrency and sequentiality analyses of cyber-physical systems specified by safe Petri nets," *IEEE Access*, vol. 7, pp. 13510–13522, 2019.

[80] A. Cheng, J. Esparza, and J. Palsberg, "Complexity results for 1-safe nets," *Theoretical Computer Science*, vol. 147, pp. 117–136, 1995.

[81] A. Giua and X. Xie, "Control of safe ordinary Petri nets using unfolding," *Discrete Event Dyn Syst*, vol. 15, pp. 349–373, 2005.

[82] A. Dideban and H. Alla, "Reduction of constraints for controller synthesis based on safe Petri nets," *Automatica*, vol. 44, pp. 1697–1706, 2008.

[83] J. Cortadella, M. Kishinevsky, L. Lavagno, and A. Yakovlev, "Deriving Petri nets from finite transition systems," *IEEE Trans. Comput.*, vol. 47, pp. 859–882, 1998.

[84] E. Fabre, "On the construction of pullbacks for safe Petri nets," in *Petri Nets and Other Models of Concurrency - ICATPN 2006,* S. Donatelli and P. S. Thiagarajan, Eds., Lecture Notes in Computer Science, Berlin, Heidelberg: Springer, pp. 166–180, 2006.

[85] E. Best and H. Wimmel, "Reducing k-safe Petri nets to pomset-equivalent 1-safe petri nets," in *Application and Theory of Petri Nets 2000,* M. Nielsen and D. Simpson, Eds., Lecture Notes in Computer Science, Berlin, Heidelberg: Springer, pp. 63–82, 2000.

[86] J. Esparza, R. Ledesma-Garza, R. Majumdar, P. Meyer, and F. Niksic, "An SMT-based approach to coverability analysis," in *Computer Aided Verification,* A. Biere and R. Bloem, Eds., Lecture Notes in Computer Science, Cham: Springer International Publishing, pp. 603–619, 2014.

CHAPTER 4

Methods of Analyzing Boundedness and Safeness

Abstract: In this chapter, novel ideas for possible improvement of Petri nets analysis methods are presented. The proposed algorithms are not only dedicated to larger Petri nets, where, as expounded in detail in the previous chapters, exact methods cannot compute in the assumed time. For larger Petri nets, that is, nets with several and more places, the described algorithms do not provide a perfect solution for all cases because, as discussed earlier, the problems are of exponential character. However, the recent ideas presented in this chapter will certainly support concurrent control designers in the field of modeling and analyzing Petri nets.

Keywords: Analysis, Boundedness, Coverability graph, Invariants, Safeness.

4.1 BOUNDEDNESS

In the proposed methods for boundedness, we focus on analyses based on reachability graphs as well as on a linear algebra approach. Each of these two approaches finds its application, as both have their own benefits and limitations. As discussed in the previous chapter, these problems possess exponential computational complexity in the general class of Petri nets [1-3]. Therefore, the proposed algorithms are geared toward efficiency and effectiveness. Efficiency is understood as obtaining a response to a given method within the assumed time. And the effectiveness is reflected in the accuracy of the result achieved. Detailed research of the proposed methods in terms of their efficiency and effectiveness is presented in Chapter 5.

4.1.1 A Method Based on Reachability Graphs

As explained in the previous chapter, boundedness analyses can be conducted by reviewing reachability graphs. This solution has an obvious disadvantage, as it requires examining all the Petri net markings, and this results in exponential computational complexity. Even for relatively small Petri nets, the current tools for testing boundedness based on state-space analysis may not yield results within the assumed time, that is, they are not efficient. Some popular computer-aided tools, together with their limitations, were discussed in Chapter 3. Upon constructing a finite graph of the reachable states of a given Petri net according to the reference algorithm shown in Algorithm 3.1, we can see that if the conditions of Equation 4.1 are met, a symbol ω is inserted in a place that will accumulate tokens indefinitely.

Marcin Wojnakowski

$$\begin{aligned} &\exists \mathbf{M_r} \text{ on the path from } \mathbf{M_0} \text{ that} \\ &\qquad \forall p \in P, M_s(p) \geq M_r(p)\ , \\ &\Longrightarrow \forall p \in P, M_s(p) > M_r(p) : M_s(p) \leftarrow \omega \end{aligned} \tag{4.1}$$

where $\mathbf{M_r} \neq \mathbf{M_s}$ and $\mathbf{M_r}$ are any reachable markings on the graph from the initial state to the current analyzed marking $\mathbf{M_s}$. The authors used this technique to keep the graph finite even for unbounded Petri nets. We shall apply this to analyze the property of boundedness, and during the construction of the graph, we shall interrupt its operation when such a place is found. It can be noticed that the proposed algorithm finds its special application for Petri nets that are unbounded, because it is not required to explore all markings of a Petri net to prove its unboundedness. The novel algorithm can work best in cases where a Petri net is suspected of being unbounded. For example, the designer has prepared a specification of a system based on a Petri net and wishes to check that he did not make any mistakes by creating an unbounded net, which would lead to an indefinite number of system states and its unpredictable behavior. The novel solution presented here will also coexist with other methods proposed further, *e.g.*, a method on a reduced row matrix, which is dedicated to rapid initial classification of boundedness.

A method for determining boundedness based on the introduced idea is presented in Algorithm 4.1. The novel solution aims mainly at prompt denial of the fact that a Petri net is bounded. The word *prompt* suggests that there is no need to construct a complete reachability graph. The option of analyzing boundedness already at the level of building state space and not after its completion, as is usually the case with other tools, is an indisputable advantage of this proposal. Common-use tools are mainly dedicated to general analysis. Thus, for example, first the state space ofa Petri net is constructed and then a certain property of the net is examined, such as boundedness, safeness or other. On the other hand, the introduced solution is restricted directly to studying the property of boundedness. The construction of the reachability graph is done on lines 2–18 by means of Algorithm 3.1. If a place is found that will keep tokens infinitely (lines 6–8), it is a place where the number of tokens in the current building state $\mathbf{M_s}$ is greater than the number of tokens for the same place in marking $\mathbf{M_r}$. $\mathbf{M_r}$ is any marking from the initial state $\mathbf{M_0}$ to $\mathbf{M_s}$, excluding that for all places; $M_s\ (p) \geq M_r\ (p)$ is satisfied. Then, such a place is marked by ω in the marking $\mathbf{M_s}$ on the graph. Next, the algorithm immediately terminates the graph construction (line 10) with a result of unboundedness (line 22). Otherwise, all states are acquired, and the net is bounded. Therefore, the proposed method is aimed at potentially unbounded Petri nets and perfectly combines with other algorithms proposed in this chapter. Details of effectiveness and efficiency and their applications will be more broadly discussed in Chapter 5.

Example 4.1 (Boundedness Analysis on a Coverability Graph) : Below, Algorithm 4.1 is presented step by step on the example of the Petri net shown in Fig. (**3.1**). The input to the algorithm is the Petri net $N_3 = (P_3, T_3, F_3, M_0)$ and the expected result is a text output: net N_3 is unbounded. At the beginning, let $\mathbf{M}$ be equal to the initial state $\mathbf{M_0}$ = [1 0 0 0 0] (line 1). Next, there is a possibility of obtaining a new state (line 2), thus we fire all the enabled transitions in the current state one by one (line 3). At this stage, it is one enabled transition t_1, after that, we obtain a new state $\mathbf{M_s} = \mathbf{M_1}$ = [0 1 0 0 0] (line 4). Then, the method checks whether the condition in Equation 4.1 is true (line 6), *i.e.*, there is such a state $\mathbf{M_r}$ reachable from the initial marking that $M_s(p) \geq M_r(p)$ for all p places. No such state is found at present. We add the current state to the collection of markings (line 13), and the fired transition to the set of fired transitions (line 14). Further, the algorithm searches for another state, *i.e.*, the next iteration (line 2). The new state can be obtained by firing the transition t_2 (line 3), so that $\mathbf{M_s} = \mathbf{M_2}$ = [1 0 1 0 0]. Again, the checker (line 6) inspects a state that satisfies the condition in Equation 4.1. Such a state $\mathbf{M_r}$ exists on the path from $\mathbf{M_0}$ to $\mathbf{M_{s=2}}$. Therefore, for any place such that $M_{s=2}(p) > M_{r=0}(p)$, an ω can be assigned (lines 7–8). The place p_3 is a place where the tokens will accumulate indefinitely since $M_2(3) > M_0(3)$, *i.e.*, the place is unbounded. It can be noticed that loop of firings $\{t_1, t_2\}$ holds tokens in the place p_3. Finally, the algorithm finishes searching for new states (line 10) and informs that net N_3 is not bounded (line 22) as expected.

Note that the algorithm for the Petri net N_3 in Fig. (**3.1**) terminates with the correct result only after two states are obtained from the initial marking. In the case of the reference method, it is necessary to build an entire reachability graph; in this example, it consists of eight distinct markings. In the worst case, such as a bounded Petri net, the proposed algorithm requires the preparation of a full graph. However, as will be shown by the experimental results presented in Chapter 5, the operating time has been optimized in practical application. The initial suspicion of unboundedness of a Petri net is a principal element, for which a novel solution described in Section 4.1.3 has been prepared.

```
Data: A Petri net N = (P, T, F, M0)
Result: The net N is bounded or unbounded
M ← an initial marking M0;
while exists a new marking that can be reachable from M do
    for each enabledtransition t at M do
        Ms ← a marking after t fired from M;
        if Ms ∉ 𝕄 then
            if exists marking Mr on the path from M0 that ∀p ∈ P, Ms(p) ≥ Mr(p)
             then
                for each place p ∈ Pt hat Ms(p) > Mr (p) do
                    Ms(p) ← ω;
                    is_bounded ← false;
                    break;
                end
            end
            𝕄 ← 𝕄 ∪ {Ms};
            Σ ← Σ ∪ {(M, Ms)};
            M ← Ms;
        end
    end
end
if is_bounded then
    print "net N is bounded";
else
    print "net N is unbounded";
end
```

Algorithm 4.1: The proposed boundedness analysis on a coverability graph.

4.1.2 A Method Based on Invariants

This subsection introduces an idea for a novel technique to analyze boundedness. The proposed method is based on computation of place invariants. It has been shown that place invariant coverage guarantees boundedness of a Petri net. The reference algorithm proposed by Martinez-Silva searches for all minimal invariants. This approach was adopted by Algorithm 3.5 to check coverage and thus analyze boundedness. However, as outlined at length in Chapter 3, theoretical computational complexity is exponential, which stands that the method may not be completed within the assumed time even for relatively small nets. Detailed experimental results are discussed in Chapter 5. It can be noticed that for the analysis of boundedness, it is enough to calculate as many place invariants that cover a Petri net as possible. Therefore, the proposal uses the Martinez-Silva algorithm and modifies it to interrupt its operation once the coverage is achieved.

Firstly, we will show the key steps of the algorithm, together with their description. Subsequently, the benefits and limitations of the solution will be discussed.

The proposed Algorithm 4.2 involves the linear algebra technique. The method is based on the Martinez-Silva computation algorithm given in Algorithm 3.3 as stated above. However, the novel solution does not require calculation of all place invariants in a Petri net. The method performs as follows. At the beginning, the unit matrix Q of matrices, identity D, and transposed incidence matrix A^{T} of the Petri net is formed (lines 1–4). In the next stage, the algorithm searches for place invariants by manipulating the matrix Q (lines 7–9). The subsequent transitions of the Petri net are examined to zeros (filling all elements with zero value) matrix A^{T}. In the meantime, matrix D saves partially obtained invariants. If all entries of any row in A^{T} are equal to zero, then the correct p-invariant can be formed from matrix D (lines 10–12). The algorithm verifies the presence of new invariants at each step and adds them to the set C (line 13). The method terminates once the obtained place invariants have covered all places of the net (lines 14–15). The place invariant coverage implies that each place refers to a non-zero place related entry of at least one invariant.

Example 4.2 (Boundedness Analysis Based on Coverage) : Below in Fig. (**2.2**), boundedness with the proposed solution of the mulit-robot system specified by the Petri net N_1 is analyzed. According to Algorithm 4.2, initially the unit matrix $Q = [D| A^{T}]$ is formed (line 4) as shown in Equation 4.2.

In the next stage, the matrix Q is transformed (lines 7–9), while the subsequent transitions are examined (line 6). Rows of A^{T} are zeroed (marked in blue) (line 7). They are referred to the place invariants formed in the matrix D (marked in green). Calculated p-invariants are added to the set I (line 12). In the selected example, only three transitions out of six in total are required to be processed to obtain the place invariant coverage. Equation 4.3 describes the matrix Q after examination

Data: A Petri net $N = (P, T, F, \mathbf{M_0})$
Result: The net N is bounded or uncovered
1 $D \leftarrow$ an identity matrix of size $|P|$;
2 $A \leftarrow$ an incidence matrix of N;
3 $A^{\mathrm{T}} \leftarrow$ `Transpose Matrix` (A);
4 $Q \leftarrow$ union matrix of D and A^{T};
5 $C \leftarrow \emptyset$;
6 **for each** *column j of matrix* A^T **do**
7   find row pairs that their sum is equal to 0 and add it to Q;
8   delete from Q rows that intersection with j-th column is not equal to 0;
9   reduce redundant rows of Q;
10   **for each** *row i of matrix* A^T **do**
11   **if** $i = 0$ **then**
12             $I \leftarrow$ a place invariant that refers to the row i in D;
13             $C \leftarrow C \cup \{I\}$;
14   **if** C *covers all places in the net N* **then**
15   is_covered $\leftarrow$ true;
16   break;
17   **end**
18   **end**
19   **end**
20 **end**
21 **if** *is_covered* **then**
22   print "net N is bounded";
23 **else**
24   print "net N is uncovered by place invariants";
25 **end**

Algorithm 4.2: The proposed boundedness analysis based on coverage.

of the third transition *i.e.*, behind three iterations of the method, four p-invariants were computed. Clearly, invariants from the set I cover all places in the Petri net (lines 14–15). Note that every place $p \in \{p_1, \cdots, p_9\}$ has a corresponding non-zero entry in at least one p-invariant. The procedure terminates its execution (line 16) with the result that the Petri net N_1 is covered by place invariants (line 22), hence it is bounded.

$$
Q = \begin{array}{c} \begin{array}{ccccccccc|cccccc} p_1 & p_2 & p_3 & p_4 & p_5 & p_6 & p_7 & p_8 & p_9 & t_1 & t_2 & t_3 & t_4 & t_5 & t_6 \end{array} \\ \left[\begin{array}{ccccccccc|cccccc}
1 & 0 & 0 & 0 & 0 & 0 & 0 & 0 & 0 & -1 & 0 & 0 & 0 & 1 & 0 \\
0 & 1 & 0 & 0 & 0 & 0 & 0 & 0 & 0 & 1 & -1 & 0 & 0 & 0 & 0 \\
0 & 0 & 1 & 0 & 0 & 0 & 0 & 0 & 0 & 0 & 1 & 0 & 0 & -1 & 0 \\
0 & 0 & 0 & 1 & 0 & 0 & 0 & 0 & 0 & 0 & 0 & -1 & 0 & 0 & 1 \\
0 & 0 & 0 & 0 & 1 & 0 & 0 & 0 & 0 & 0 & 0 & 1 & -1 & 0 & 0 \\
0 & 0 & 0 & 0 & 0 & 1 & 0 & 0 & 0 & 0 & 0 & 0 & 1 & 0 & -1 \\
0 & 0 & 0 & 0 & 0 & 0 & 1 & 0 & 0 & 0 & -1 & 0 & -1 & 1 & 1 \\
0 & 0 & 0 & 0 & 0 & 0 & 0 & 1 & 0 & 0 & -1 & 0 & 1 & 0 & 0 \\
0 & 0 & 0 & 0 & 0 & 0 & 0 & 0 & 1 & 0 & 1 & 0 & -1 & 0 & 0
\end{array}\right] \end{array} \tag{4.2}
$$

$$
Q = \begin{array}{c} \begin{array}{ccccccccc|cccccc} p_1 & p_2 & p_3 & p_4 & p_5 & p_6 & p_7 & p_8 & p_9 & t_1 & t_2 & t_3 & t_4 & t_5 & t_6 \end{array} \\ \left[\begin{array}{ccccccccc|cccccc}
0 & 0 & 0 & 0 & 0 & 0 & 0 & 1 & 1 & 0 & 0 & 0 & 0 & 0 & 0 \\
1 & 1 & 1 & 0 & 0 & 0 & 0 & 0 & 0 & 0 & 0 & 0 & 0 & 0 & 0 \\
0 & 0 & 1 & 0 & 0 & 1 & 1 & 0 & 0 & 0 & 0 & 0 & 0 & 0 & 0 \\
0 & 0 & 2 & 0 & 0 & 0 & 1 & 1 & 0 & 0 & 0 & 0 & 0 & -1 & 1 \\
1 & 1 & 0 & 0 & 0 & 1 & 0 & 0 & 1 & 0 & 0 & 0 & 0 & 1 & -1 \\
0 & 0 & 0 & 1 & 1 & 1 & 0 & 0 & 0 & 0 & 0 & 0 & 0 & 0 & 0 \\
0 & 0 & 1 & 1 & 1 & 0 & 0 & 1 & 0 & 0 & 0 & 0 & 0 & -1 & 1
\end{array}\right] \end{array} \tag{4.3}
$$

Since a Petri net is covered by place invariants, the Petri net is bounded. We can notice that not all invariants are required to compute, but only as many as it is needed to cover the Petri net contrary to the typical method given in Algorithm 3.5. On the other hand, in the worst case, the algorithm is obliged to compute all the invariants. And also, if the net is not covered, the boundedness is not determined. Therefore, the novel algorithm suggests finding its best application for covered Petri nets. Analyzing the boundedness of an uncovered Petri net in this way does not bring any changes. Hence, at early stage, it rises the need to quickly estimate whether a given Petri net could potentially be covered or uncovered by place invariants. This is the purpose of the next proposed method.

4.1.3 A Method Based on Reduced Row Echelon Form

The next proposed method uses common linear algebra matrix operations like Gauss-Jordan elimination [4] to solve linear systems. The search for invariants can be checked to solve the homogenous linear system $A\mathbf{x}$, where A is an incidence matrix of a Petri net. Section 2.8 states that one of possible methodologies to solve such a system is to adopt a matrix in the reduced row echelon form that can be obtained after Gauss-Jordan elimination. We also know that a Petri net covered by p-invariants is bounded (c.f. Chapter 3). Combining it, we propose a novel algorithm to support designers in checking the boundedness of Petri nets. It is particularly applicable in the preliminary determination of whether a Petri net cannot be covered by place invariants. The presented technique does not permit obtaining all-place invariants. Moreover, it operates in polynomial time.

Proposition 4.1 :

A Petri net cannot be covered by place invariants if there exists at least one row containing only a leading one in a matrix B, where the matrix B is in the reduced row echelon form of the incidence matrix A of the Petri net.

Proof: If there exists a row that contains only a leading one in matrix B, according to Def. 2.42 (solving linear systems $A\mathbf{x} = 0$ with application of matrix in the reduced row echelon form)

$$p_i = 0\,, \tag{4.4}$$

where i is the i-th column index of matrix B that correspond to the place i of a Petri net. Hence, place i is never covered since i-th entry of any $\mathbf{x}$ always equals to 0. Thus, the Petri net cannot be covered by p-invariants.

Proposition 4.2 :

A Petri net cannot be covered by place invariants if there exists at least one row containing only ones after a leading one in a matrix B, where the matrix B is in the reduced row echelon form of the incidence matrix A of the Petri net.

Proof: If there exists a row that contains only leading one and ones in matrix B, according to Def. 2.42 (solving linear systems $A\mathbf{x} = 0$ with application of matrix in reduced row echelon form)

$$p_i = -p_{j1} + (-p_{j2}) + \cdot\cdot\cdot + (-p_{jk})\,, \tag{4.5}$$

where i is the i-th column index with leading one and j -th k columns with ones. Hence, $p_i \leq 0$; place i is never covered since i-th entry of any $\mathbf{x}$ is always equal or less to 0. Thus, the Petri net cannot be covered by p-invariants.

> **Important**

It is worth noting that if a Petri net contains any transitions that do not have input places, the behavior of the methods is unexpected. A transition that only has output places generates tokens ad infinitum. Such nets are obliviously unbounded.

To disprove that a Petri net can be covered by place invariants, the proposed polynomial-time method is divided into following steps presented in Algorithm 4.3. The idea of the proposed method is to perform Gauss-Jordan elimination in order to obtain the reduced row echelon form of the incidence matrix (lines 2–11). Then, we test conditions that disprove that a Petri net can be covered by place invariants. Proposition 4.1 is tested on lines 7–10 of the algorithm, and Proposition 4.2 is assessed on lines 11–14. If a Petri net is not p-invariant coverable, the proposed algorithm proves it in an efficient way *i.e.*, polynomial time.

> **Important**

We know that a place invariant cover determines boundedness, however a Petri net that is not covered cannot be claimed to be unbounded.

On the other hand, we do not know whether a Petri net is covered (bounded) if none of the proposed condition appears. It is the main bottleneck of the novel solution.

However, the experimental results show, and are subjected to detailed analyses in Chapter 5, that if a Petri net is not marked as uncovered by the algorithm, it is likely to be covered, *i.e.*, bounded. Moreover, the information about the non-coverage of a Petri net by place invariants can be used to select methods for further analyses of boundedness. For not covered Petri nets, it is pointless to use solutions based on computation of invariants and then to use algorithms, for example, on reachability graphs instead.

```
Data: A Petri net N = (P, T, F, M0)
A ← an incidence matrix of N;
Function Perform Gauss-Jordan Elimination(A):
    B ← A;
    swap rows so that all rows with all zero entries are on the bottom;
    repeat
        swap rows so that the row with the largest, leftmost nonzero entry is on top;
        multiply the top row by a scalar so that top row's leading entry becomes 1;
        add/subtract multiples of the top row to the other rows so that all other
          entries in the column containing the top row's leading entry are all zero;
        swap rows so that the leading entry of each nonzero row is to the right of the
          leading entry of the row above it;
    until all the leading entries are 1;
    return B;
B ← Perform Gauss-Jordan Elimination(A);
for each row i of the matrix B do
    if i has only a leading one then
        is_not_covered ← true;
        break;
    end
    if i has only ones then
        is_not_covered ← true;
        break;
    end
end
if is_not_covered then
    print "net N is not covered by p-invariants";
else
    print "net N might be covered by place invariants, therefore it may be bounded.";
end
```

Algorithm 4.3: A proposed invariants coverage disproving procedure.

Example 4.3 (Non-p-invariant Covered) : The next example shows a Petri net N_6 [1] in Fig. (**4.1**) that is bounded and not covered by p-invariants. The incidence matrix of the Petri net N_6 after Gauss-Jordan elimination in the reduced row echelon form is presented in Equation 4.6.

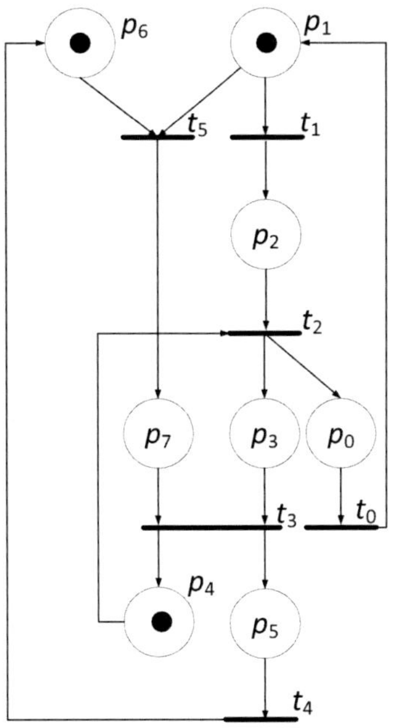

Fig. (4.1). Non-p-invariant covered but a bounded Petri net N_6.

The reduced row echelon form has three rows with just a leading one that $p_1 = 0$, $p_0 = 0$, and $p_2 = 0$, thus places p_1, p_0, and p_2 are not covered. Whereas, the reachability graph of the Petri net shown in Fig. (**4.2**) confirms that the net is bounded, where $k = 1$. Table **4.1** shows markings of every reachable state of the Petri net N_6. An example of a place invariant uncovered but a bounded Petri net is described to indicate that an uncovered net does not imply unboundedness. However, it should be emphasized that the occurrence of such Petri nets is relatively rare. Even though experimental verification is conducted on hundreds various Petri nets in Chapter 5, the percentage of such cases is negligible.

[1] Available online: hippo.uz.zgora.pl, *cp2* benchmark (accessed on November 30, 2024)

Table 4.1 Description of each reachable marking in N_6.

-	p_1	p_0	p_2	p_3	p_4	p_7	p_5	p_6
M_0	1	0	0	0	1	0	0	1
M_1	0	0	0	0	1	1	0	0
M_2	0	0	1	0	1	0	0	1
M_3	0	1	0	1	0	0	0	1
M_4	1	0	0	1	0	0	0	1
M_5	0	0	0	1	0	1	0	0
M_6	0	0	1	1	0	0	0	1
M_7	0	0	0	0	1	0	1	0
M_8	0	0	0	0	1	0	0	1

$$B = \begin{array}{c} \begin{array}{cccccccc} p_1 & p_0 & p_2 & p_3 & p_4 & p_7 & p_5 & p_6 \end{array} \\ \begin{bmatrix} 1 & 0 & 0 & 0 & 0 & 0 & 0 & 0 \\ 0 & 1 & 0 & 0 & 0 & 0 & 0 & 0 \\ 0 & 0 & 1 & 0 & 0 & 0 & 0 & 0 \\ 0 & 0 & 0 & 1 & -1 & 0 & 0 & 0 \\ 0 & 0 & 0 & 0 & 0 & 1 & 0 & -1 \\ 0 & 0 & 0 & 0 & 0 & 0 & 1 & -1 \end{bmatrix} \end{array} \qquad (4.6)$$

Gauss-Jordan elimination constitutes a core of the proposed method. The run-time of the algorithm depends mainly on a chosen Gauss-Jordan implementation. Nowadays, extremely fast and efficient variants of Gaussian Elimination are available [5-7], for instance GPU-based [8]. Most mathematical tools (*e.g.*, MATLAB [2]) offer quick numerical methods [3].

The computational complexity of the proposed algorithm is bounded by $O(|T|^2|P|)$, where $|T|$ is the number of transitions and $|P|$ is the number of places of a Petri net. The algorithm is divided into two parts. At first, transformation of the incidence matrix to the reduced row echelon form is executed at most $|T|^2|P|$ times. The second part with the outer loop and nested loops is executed at most $|T||P|$ times. The presented method shows that a Petri net that is non-covered by place invariants can be detected in polynomial time.

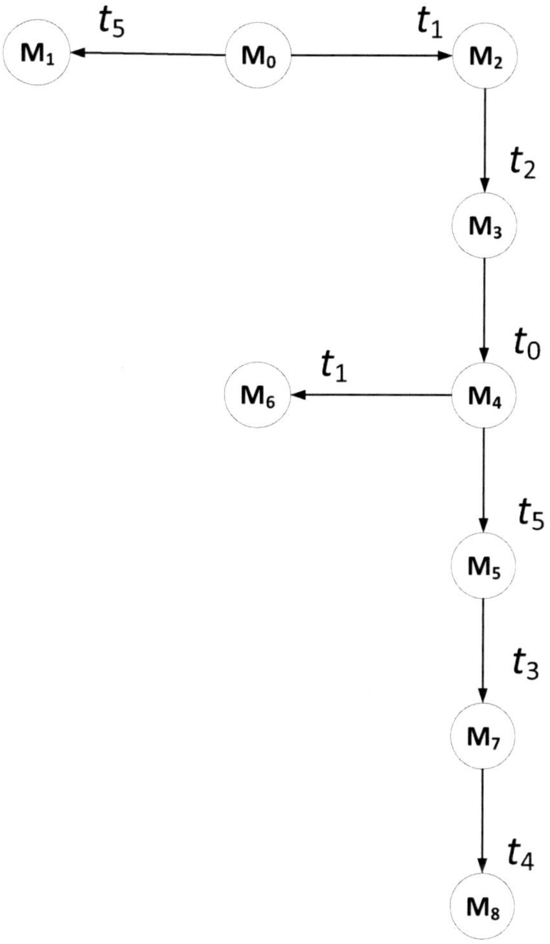

Fig. (4.2). The state space of Petri net N_6.

---

[2] MATLAB combines a desktop environment tuned for iterative analysis and design processes with a programming language that expresses matrix and array mathematics directly. More information available online: https://www.mathworks.com/products/matlab.html (accessed on November 30, 2024)

[3] For instance, *rref* function in MATLAB. More information available online: https://www.mathworks.com/help/matlab/ref/rref.html (accessed on November 30, 2024)

4.2 SAFENESS

In the proposed methods for checking safeness, we shall utilize two methodologies for the computation of invariants with the formation of state machine components: exploration of reachability graphs or the linear algebra approach. Similar techniques, as for the verification of boundedness, are proposed for the analysis of safeness to improve time efficiency compared to existing methods that maintain effectiveness (correctness of results). As explained in the previous chapter, state reachability and the problem of searching for all minimal invariants are time-exponential, in general. Hence, the novel proposed algorithms find their best application in particular situations. More details on the effectiveness and efficiency of the proposed methods will be introduced in Chapter 5.

4.2.1 A Method Based on Reachability Graphs

Safeness can be analyzed by means of exploration of reachability graphs. The construction of a full graph involves exponential time computational complexity, which we explained in detail in the previous chapter. It can be noticed that during the construction of the graph, the number of tokens for particular places is determined by subsequent markings. Hence, during the building of the state space, it is possible to check whether a Petri net can be safe, *i.e.*, there is no such state in which even one place has more than one token. Therefore, the algorithm can be terminated without the need to designate all states the moment a place contains more than one token. Moreover, in the case of unbounded place detection, such a Petri net cannot also be safe, because if the condition in Equation 4.1 is satisfied, then there exists a place that accumulates tokens indefinitely. Thus, the execution of the method may also be interrupted.

In the following, a novel algorithm based on the construction of a coverability graph for analysis of safeness of concurrent control systems specified by Petri nets is presented. The core idea of this method during the construction of the graph is to verify whether there will be a marking that contains more than one token in some place. In such a case, we can terminate the operation of the algorithm, claiming that the Petri net is unsafe.

Algorithm 4.4 presents a solution that implements the described idea. The proposed method may determine that a Petri net is not safe in the process of building state spaces. Of course, if a Petri net is safe, it is required to construct the entire coverability graph. Therefore, if a Petri net is suspected of being unbounded, *e.g.*, by a method based on a reduced row echelon form, it may not be safe, since any unbounded net is also unsafe by definition. In that case, the use of this proposed algorithm may be the most advisable.

Initially, the proposed algorithm requests a Petri net *N* as an input. Then, the initial marking $\mathbf{M_0}$ is assigned to the variable $\mathbf{M}$ (line 1). In the next step, a loop begins, which works until a new state cannot be designated (line 2). In the loop, after subsequent firing of the enabled transitions (line 3), a new marking is determined and assigned to the variable $\mathbf{Ms}$ (line 4). If there is a place in this marking where there is more than one token (line 7), then the loop is broken (line 9) and the algorithm terminates with the result the Petri net *N* being unsafe (line 27). Similarly, the algorithm terminates if an ω is assigned to a place, *i.e.*, a place that will accumulate tokens indefinitely (line 14). Marking a place as unbounded is based on a condition from the Equation 4.1 (see Section 4.1 for more details). Otherwise, if no unsafe place occurs, the algorithm continues by firing the next enabled transition (line 3) and designating a new place until there is no new unique marking (line 23). The Petri net *N* is safe (line 25) when each place in any reachable marking has at most one token.

4.2.2 A Method Based on Invariants

Another method introduces an algorithm for analyzing safeness based on the computing of place invariants that form correct state machine components. It is claimed (q.v. Section 3.5) that a Petri net covered by SM-components is safe. As already mentioned in Section 3.3, a Petri net covered by place invariants is bounded, and if it is also covered by SMCs, it is safe. In view of the above, the proposed algorithm for the analysis of boundedness is extended to check whether the obtained invariant composes the correct SMCs. That is, whether a SM-net has exactly one token in the initial marking. When all the state machine components cover the Petri net, *i.e.*, every place of the Petri net is a part of at least one SMC, then the algorithm stops its operation with the result of the Petri net being safe. The method is first described, after which it is explained by its profits and limitations. The novel solution can be presented by the steps shown in Algorithm 4.5.

```
Data: A Petri net N = (P, T, F, M0)
Result: The net N is safe or unsafe
M ← an initial marking M0;
while exists a new marking that can be reachable from M do
    for each enabled transition t at M do
        Ms ← a marking after t fired from M;
        if Ms ∉ 𝕄 then
            for each place p in Ms do
                if Ms(p) > 1 then
                    is_safe ← false;
                    break;
                end
            end
            if exists Mr on the path from M0 that ∀p ∈ P, Ms(p) ≥ Mr(p) then
                if ∀p ∈ P, Ms(p) > Mr(p) then
                    Ms(p) ← ω;
                    is_safe ← false;
                    break;
                end
            end
            𝕄 ← 𝕄 ∪ {Ms};
            Σ ← Σ ∪ {(M, Ms)};
            M ← Ms;
        end
    end
end
if is_safe then
    print "net N is safe";
else
    print "net N is unsafe";
end
```

Algorithm 4.4: The proposed safeness analysis on a coverability graph.

Firstly, initialization is performed to read an incidence matrix of a Petri net N. In addition, the necessary variables are set: D — an identity matrix, A^{T} — the transposed incidence matrix, and C —— an empty set for the cover of SMC (lines 1–4).

```
   Data: A Petri net N = (P, T, F, M0)
1  A ← an incidence matrix of N;
2  D ← an identity matrix of size |P|;
3  A^T ← Transpose Matrix (A);
4  C ← ∅;
5  Q ← union matrix of D and A^T;
6  for each column j of matrix A^T do
7  |   find row pairs that their sum is equal to 0 and add it to Q;
8  |   delete from Q rows that intersection with j-th column is not equal to 0;
9  |   reduce redundant rows of Q;
10 |   for each row i of matrix A^T do
11 |   |   if i = 0 then
12 |   |   |   I ← a place invariant that refers to the row i in D;
13 |   |   |   if I forms a proper SMC (contains exactly one token in M0) then
14 |   |   |   |   C ← C ∪ {I};
15 |   |   |   end
16 |   |   end
17 |   |   if C covers all places then
18 |   |   |   is_SMCs_covered ← true;
19 |   |   |   break;
20 |   |   end
21 |   end
22 end
23 if is_SMCs_covered then
24 |   print "net N is safe";
25 else
26 |   print "net N is uncovered by SMCs";
27 end
```

Algorithm 4.5: The proposed SMCs cover-based boundedness analysis.

In the next stage, the unit matrix Q is formed as a conjunction of the identity matrix D and the transposed incidence matrix of the Petri net system A^{T} (line 5). Note that D will hold the place invariants during examination of subsequent transitions (lines 7–9). If a row in A^{T} is zeroed (all entries are equal to zero), then the correct p-invariant can be read from the corresponding row in matrix D. Checking for the state machine cover is a key objective of the algorithm. This step is executed for each transition until the SM-cover is found. At each stage, the algorithm verifies whether the obtained place invariants I form a proper state machine component (lines 12 & 13). A correct S-component contains exactly one token in the initially marked place. Then, SM-component is added to set C (line 14).

Simultaneously, the existence of the SMC cover, understood as state machine components that cover all places of the net, is analyzed. The execution of the method is terminated when the coverage is gained (lines 17–20) with the result that the net is safe (lines 23–24).

Otherwise, all state machine components are obtained and the algorithm informs about the absence of coverage. If a Petri net is not covered by S-components, the safeness is not determined by this algorithm because a not-covered net may be both safe or unsafe. Moreover, the result may not be obtained in reasonable time as all p-invariants are required to compute.

Example 4.4 (Safeness Analysis Based on Coverage) : Below, the proposed method is presented on the example of the Petri net N_1, which models the operation of the multi-robot shown in Fig. (**2.2**). The first step of the algorithm is to input the Petri net as the incident matrix A_1 shown in Equation 2.16. Then the incidence matrix is transposed and united with the identity matrix D. That matrix is presented in Equation 4.2. Successively, for each column j in the matrix A^T, such pairs of rows are sought whose sum is equal to zero. These rows are added to the Q matrix, while redundant rows and those whose intersection with j column is not equal to zero are removed from the matrix Q. In the next step, we check whether any row in the matrix A^T has been zeroed, if so, the corresponding row in the matrix D forms a place invariant. After examination of three transitions (three iterations: $j = \{1, 2, 3\}$), the matrix Q is given in Equation 4.3. In the presented example, there are four such place invariants in the set I:

Simultaneously, we verify whether the obtained invariant composes the correct state machine component, that is, whether it has exactly one token in the place marked initially (marked in blue), if so, SMC is added to set C. After each S-component obtained, the SMC-cover is checked. The coverage is established after three of six transitions have been examined. The loop is broken and the algorithm ends with the information that the Petri net N_1 is safe.

$$\begin{aligned} I(1) &= [0\,0\,0\,0\,0\,0\,0\,1\,1] \\ I(2) &= [1\,1\,1\,0\,0\,0\,0\,0\,0] \\ I(3) &= [0\,0\,1\,0\,0\,1\,1\,0\,0] \\ I(4) &= [0\,0\,0\,1\,1\,1\,0\,0\,0]\,. \end{aligned} \tag{4.7}$$

CONCLUSION

The novel algorithms proposed are meant to bridge the existing gap. The described proposals for the analysis of boundedness and safeness are based on two approaches: state-space exploration and linear algebra (invariant computation). Three new methods of verification of boundedness and two of safeness have been proposed. The problems of constructing reachability graphs and calculating all minimum invariants are exponential with respect to time. Hence, the novel proposed methods bring about improvements in special cases. A detailed analysis of the efficiency and effectiveness of the solution is presented in Chapter 5. It is worth adding that the method that applies an incidence matrix in the reduced row echelon form matrix can reject place-invariant coverage in polynomial time or determine with high probability that it is covered, that is, potentially bounded. The presented solutions terminate the operation of the algorithms regardless of the conditions that confirm or reject the properties of the Petri net being met. Some bottlenecks of the novel methods are also described in the next chapter with two main limitations being: a Petri net that is not covered by place invariants is not always unbounded, and a net that is uncovered by S-components can also be safe.

REFERENCES

[1] R. J. Lipton, "The reachability problem requires exponential space," in *Technical Report 62*, Yale University, 1976.

[2] K. Barkaoui and M. Minoux, "A polynomial-time graph algorithm to decide liveness of some basic classes of bounded Petri nets," in *Application and Theory of Petri Nets 1992*. K. Jensen, Ed., Lecture Notes in Computer Science, Berlin, Heidelberg: Springer, pp. 62–75, 1992.

[3] N. D. Jones, L. H. Landweber, and Y. Edmund Lien, "Complexity of some problems in Petri nets," *Theor. Comput. Sci.*, vol. 4, pp. 277–299, 1977.

[4] A. T. Yuster and T. Yuster, "The reduced row echelon form of a matrix is unique: A simple proof," *Mathematics Magazine*, 1984.

[5] P. S. Stanimirović and M. D. Petković , "Gauss–Jordan elimination method for computing outer inverses," *Appl. Math. Comput.*, vol. 219, pp. 4667–4679, 2013.

[6] G. Sharma, A. Agarwala, and B. Bhattacharya, "A fast parallel Gauss Jordan algorithm for matrix inversion using CUDA," *Comput. Struct.*, vol. 128, pp. 31–37, 2013.

[7] D. N. Diep, D. H. Giang, and N. Van Minh, "Quantum gauss-jordan elimination and simulation of accounting principles on quantum computers," *Int J Theor Phys*, vol. 56, pp. 1948–1960, 2017.

[8] N. A. Atasoy, B. Sen, and B. Selcuk, "Using gauss - Jordan elimination method with CUDA for linear circuit equation systems," *Procedia Technol.*, vol. 1, pp. 31–35, 2012.

CHAPTER 5

Experimental Verification of the Proposed Methods

Abstract: Experimental research is an essential element in developing novel algorithms and testing already available solutions for their optimal application. Each proposed algorithm should be evaluated using reference methods in terms of its accuracy and operating time. The algorithms introduced in Chapter 4 were tested experimentally. The conducted research was meant to compare the presented methods with reference algorithms. Here, we also discuss the best application of the selected algorithms in order to verify boundedness and safeness. The benchmark library used in the experiments, containing 234 Petri nets, is part of a Hippo project developed at the University of Zielona Góra. The set of test modules consists of various classes as well as features various levels of complexity of the Petri nets used for the specification of hypothetical and real-life examples of control systems.

Keywords: Analysis, Boundedness, Benchmarks, Experiments, Safeness, Verification.

5.1 INTRODUCTION

Our research focuses on inspecting the effectiveness and efficiency of the selected algorithms. As already stated in previous chapters, effectiveness is understood here as consistency of results and efficiency as time needed to achieve a result. In the case of methods based on the calculation of invariants, the amount of system memory usage is not an obstacle at this stage. Exponentiality with respect to memory is an important optimization problem within state space construction. However, since we must first deal with time complexity, memory performance tests are not part of this work. A comparison of algorithms allowed us to indicate the advantages and disadvantages of the presented methods and to indicate the best cases of their use.

In the following, the Hippo system that was used in the experimental research is introduced. The introduction is followed by a comparison of the proposed analyses

Marcin Wojnakowski

of boundedness algorithms to reference methods. Later, safeness methods are scrutinized, and finally, the summary of methods that are optimal to apply in particular cases is presented.

5.2 THE HIPPO SYSTEM

There are various tools available on the market to work with Petri nets. Some of these software are probably known only to their authors, and the others are problematic even in their installation or launch. Only some are suitable for daily use by end users, that is, designers of concurrent control systems based on Petri nets. In response to this demand, Hippo [1] [1-3] was launched in 2005, designed as a set of standalone tools for broad Petri net analysis. Initially, the components were oriented to systems based on hypergraph theory. The computer-aided design system has been significantly developed over the past several years. Now, it focuses on control systems specified by a Petri net. The most recent version of Hippo consists of a set of programs that support control system designers, including prototyping, verification, analysis, decomposition, modeling, and automatic code generation (implementation). Furthermore, since they provide a balance between optimal results and computational time, the modules allow solving the same problems using various methods.

Hippo tools include three completely different approaches to the analysis (or decomposition) of control systems. They can be used depending on one's needs. In the previous chapters, we stated that there was no perfect analytical method in the general case of Petri nets. Therefore, the following techniques are available:

- linear algebra approach,
- algorithms based on graph theory,
- algorithms based on hypergraph theory.

In the following, the methods included in Hippo are presented. Moreover, currently, the Hippo project consists of a library of 264 various Petri nets. The software

[1] The detailed description of the project and access to Petri net benchmarks can be found online: https://hippo.uz.zgora.pl (accessed on November 30, 2024)

accepts input as a PNH [2] file. The format includes an incidence matrix of a Petri net and an initial marking vector. In addition, other options can be specified. However, they are optional for analytical proposes. Hippo permits verification by classification of Petri nets, analysis of concurrency, and sequentiality relations. Furthermore, it allows the computation of place invariants and the checks of the safeness and boundedness of the Petri nets. Finally, the Hippo system can return the model in one of the following formats: Verilog (for FPGA implementation), PNG (graphical representation of the model), and XML (for PIPE).

5.3 BOUNDEDNESS

The novel algorithms for boundedness analysis have been introduced in Section 4.1. Three experiments are performed. The first compares methods based on reachability graphs. The second report exposes algorithms based on linear algebra, *i.e.*, computation of invariants of places. The sample results of the conducted research are shown in Table **5.1** (reachability graph-based algorithms) and Table **5.2** (invariant coverage-based algorithms). Furthermore, the third experiment, *i.e.*, a method to estimate which of the proposed algorithms should be used in the particular case, is shown in Table **5.3.** The full results of all experiments are attached in Appendix A. The tests are prepared using a dedicated computational server equipped with an Intel(R) Xeon(R) Gold 5220 @2.2 GHz processor and 128 GB of RAM.

First of all, a fundamental difference between methods based on browsing state space and computation of place invariants can be identified. An overview of all possible states gives the answer to whether a Petri net is bounded or unbounded. On the other hand, algorithms based on computing invariants are usually faster (with shorter run-time). In the case of analysis of invariants, we talk about their coverage of a Petri net. The coverage guarantees the boundedness of a Petri net; however, the lack of coverage does not indicate that the net cannot be bounded. Generally, such cases are rare. In our test library, there are 14 such Petri nets out of 243, which is less than 6%. Of the selected Petri nets shown in the tables, *balduzzi1* is an example of an uncovered and bounded Petri net.

Table **5.1** compiles reachability graph-based algorithms for boundedness verification. The reference method for full-graph construction presented in Algorithm 3.1 and its implementation for the purpose of experimental research is the first reference

[2] Petri Net Hippo format

Table 5.1 Sample results of experiments, state space-based algorithms.

Name	$\|P\|$	$\|T\|$	1st reference method bounded	runtime[ms]	1st proposed method bounded	runtime[ms]
coloured	4	3	true	25.564	true	17.262
cortadella1	9	6	true	29.867	true	17.477
board_game	13	13	false	21.418	false	15.285
lnet_p6n2	14	16	false	1,316.918	false	49.137
dining_philosophers	15	10	false	124.836	false	45.724
brenner1	16	12	true	51,507.596	true	49,074.995
state_space16	16	16	true	3,208.890	true	1,802.107
automation3	17	16	true	66,731	true	78.435
balduzzi1	18	16	true	31.694	true	30.205
adam1	24	12	true	57.150	true	55.183
HAN	24	40	true	100,185.767	true	50,454.547
crossroadSM_FPGA	32	12	true	79.008	true	30.463
state_space32	32	32	n/a	n/a	n/a	n/a
state_space48	48	48	n/a	n/a	n/a	n/a
cn_crr7	56	15	true	191,885.115	true	95,667.592
cn_crr10	80	21	n/a	n/a	n/a	n/a
cn_crr15	120	31	n/a	n/a	n/a	n/a
cn_crr25	200	51	n/a	n/a	n/a	n/a

method (q.v. Algorithm 3.4). It is used to compare the first proposed algorithm. The first proposal with the reduced state space (terminating if an unbounded place is located) is given in Algorithm 4.1. Both of them are based on browsing the space of possible reachable states. The marking *n/a* indicates that a result is not obtained in the assumed time, that is, an hour. The particular columns contain the following values:

- Name — the name of a Petri net (benchmark),
- $|P|$ — the number of places in the Petri net,
- $|T|$ — the number of transitions in the Petri net,
- bounded — whether the Petri net is bounded,
- runtime — the run-time of the respective methods in milliseconds,
- 1st reference method — a full reachability graph-based reference algorithm,
- 1st proposed method — a proposed reduced reachability graph-based algorithm.

Computational complexity, which is not directly dependent on the number of places and transitions in a Petri net [4], is what should be considered when comparing the algorithms of boundedness property analysis based on state space. The time needed to verify boundedness of a Petri net depends on the number of reachable states. This is illustrated in Table **5.1** by an example of a Petri net with fourteen places and sixteen transitions *lnet_p6n2*. The first reference method takes more than one second (1,316.918 ms) to obtain the result, while for the larger Petri net *crossroadSM_FPGA*, it takes just seventy-nine milliseconds.

The first proposed method is applicable to Petri nets that are unbounded since the result can be obtained quickly by interrupting the operation of the algorithm after detecting an unbounded place. This can be seen in the example of the Petri net *lnet_p6n2*, for which it takes 1.3 seconds (1,316.918 ms) to build the entire reachability graph and view all the states (the 1st reference method), while the determination of the state space and the simultaneous verification of the unboundedness places are performed almost 27 times faster (the 1st proposed method). Since the Petri net is not bounded and unbounded, the place is detected early. For bounded Petri nets, the result time is close to the reference method. Although with larger Petri nets, we can see some improvement resulting from the fact that the checking of boundedness of places occurs already during the construction of the reachability graph. Since there is no need to browse the designated state space, the absence of unbounded places means that the Petri net is bounded. The *HAN* Petri net, in which the total time of the property analysis is two times shorter compared to the reference method, is a good example of this. Hence, a solution that can estimate whether a Petri net may not be bounded at the beginning of the boundedness analysis procedure is needed. The third proposed method is dedicated to such cases. In the case of potentially unbounded Petri nets, a beneficial solution seems to choose the first proposed method. On the other hand, it is more efficient to use the other method (the next proposed method is discussed in the following paragraph) for bounded nets. Generally, state-space analysis makes it possible to answer whether a Petri net is bounded or unbounded. However, with current methods, computational complexity of the reachability graph construction makes it impossible to complete it for some even relatively small, bounded Petri nets (several places and transitions), such as *state_space32*, *state_space48*, *cn_crr10*, *cn_crr15*, or *cn_crr25*.

Table **5.2** presents research on algorithms that analyze the property of boundedness based on the verification of the coverage of the Petri net by invariants of places. The Martinez-Silva (M-S) algorithm (q.v. Algorithm 3.3) is a popular classic method for computing all minimal invariants (the full set) and is presented at length in

Table 5.2 Sample results of experiments, linear algebra-based algorithms

Name	|P|	|T |	2nd Reference Method Covered Runtime[ms]		2nd Proposed Method Covered Runtime[ms]	
coloured	4	3	true	0.045	true	0.088
cortadella1	9	6	true	0.065	true	0.124
board_game	13	13	false	0.104	false	0.237
lnet_p6n2	14	16	false	0.085	false	0.240
dining_philosophers	15	10	false	0.068	false	0.331
brenner1	16	12	true	0.095	true	0.519
state_space16	16	16	true	0.091	true	2.167
automation3	17	16	false	0.083	false	0.567
balduzzi1	18	16	false	0.107	false	0.656
adam1	24	12	true	1.805	true	1.884
HAN	24	40	true	0.102	true	1.235
crossroadSM_FPGA	32	12	true	891,400.791	true	1.449
state_space32	32	32	true	0.264	true	2.254
state_space48	48	48	true	0.375	true	8.344
cn_crr7	56	15	n/a	n/a	true	3.832
cn_crr10	80	21	n/a	n/a	true	12.782
cn_crr15	120	31	n/a	n/a	true	37.731
cn_crr25	200	51	n/a	n/a	true	206.393

Section 3.3. Here, it is an implemented analysis of boundedness of the Petri net (q.v. Algorithm 3.5). The examined method is the second algorithm proposed in Algorithm 4.2. The particular columns describe the following values:

- covered — whether the Petri net is covered by place invariants therefore bounded,
- runtime — the run-time of the respective methods in milliseconds,
- 2nd reference method — the reference classic (full invariants set) algorithm implemented toward p-invariants coverage verification,
- 2nd proposed method — the proposed algorithm based on terminating operation while the place invariants coverage is obtained (set of reduced invariants).

Table **5.2** compares the two methods based on invariant computation differently from generating a state space. The compared algorithms check the coverage of a Petri net with the invariants of places. The second reference method is a classic algorithm for obtaining all the minimal invariants in a Petri net. In our research, it was adapted to the Petri net coverage for the requested boundedness analysis. The guarantee of the correctness of the result obtained is an undoubted advantage. However, even with relatively small Petri nets, such as *crossroadSM FPGA* (32 places, 12 transitions), the time required to perform the analysis is almost 14 minutes (891,400.791 ms). And for *cn crr7* and larger nets, the run-time is more than one hour; after that time, the algorithm is terminated (the result was not found in the assumed time). In such a case, we claim that the method is inefficient.

The second proposed method, while calculating the invariants, checks whether a Petri net has already been covered by them. On the one hand, this is associated with a greater number of conditions. Thus, for exceedingly small nets, for example, *colored* or *cortadella1*, the time is slightly higher for this algorithm. On the other hand, the difference is essential for larger nets, *e.g.*, *crossroadSM FPGA* as much as 4,474 times faster. For *cn crr7* and nets with more places and transitions, there is a result, as opposed to no result with the reference method. However, the operating time on the *cn crr25* net with 200 places and 51 transitions is more than one minute. The computational complexity of this algorithm is obviously exponential, *i.e.*, in the worst case, it requires to obtain all the minimal invariants, especially when a Petri net is not covered. In such a case, the next proposed method may deliver the result in an instant.

Table **5.3** presents the results of the tests of the third proposed algorithm presented in Algorithm 4.3. To the best of our knowledge, it is currently the fastest method of an approximate examination of Petri net coverage by place invariants. Although the algorithm, as proven, works in polynomial time, it only guarantees that a Petri net is uncovered by place invariants. Otherwise, it is highly likely that a Petri net may be covered by place invariants, but this has not been confirmed. However, we have not found such a case in our extensive experimental research, which does not mean that it does not exist. When informed that a Petri net is not covered, the designer receives information that it is highly probable that the net is not properly designed. Even if it can be bounded, it is not live. In the reverse case, the designer cannot be sure of the boundedness of the Petri net. However, this may be the only solution capable of any property analysis for larger nets, *i.e.*, such with more than hundreds of places.

The third novel algorithm uses Gauss-Jordan elimination and is based on a matrix in a reduced row echelon form. It is mainly used to determine which method for further boundedness analysis to choose: the exact one, *i.e.*, the first proposal based on a reduced reachability graph or the approximate one, *i.e.*, the second proposal on the coverage of invariants (the reduced set). The algorithm indicates whether a Petri net is not covered by place invariants, hence it is probably unbounded, and then the first proposed method may effectively and efficiently check it. In addition, the rows

Table 5.3 Sample results of experiments, the reduced row echelon form-based algorithm.

Name	|*P*|	|*T*|	3rd Proposed Method Covered	Further Analysis	Runtime[ms]
coloured	4	3	possibly	2nd proposed method	0.003
cortadella1	9	6	possibly	2nd proposed method	0.003
board_game	13	13	false	1st proposed method	0.007
lnet_p6n2	14	16	false	1st proposed method	0.008
dining_philosophers	15	10	false	1st proposed method	0.006
brenner1	16	12	possibly	2nd proposed method	0.010
state_space16	16	16	possibly	2nd proposed method	0.008
automation3	17	16	false	1st proposed method	0.010
balduzzi1	18	16	false	1st proposed method	0.009
adam1	24	12	possibly	2nd proposed method	0.008
HAN	24	40	possibly	2nd proposed method	0.006
crossroadSM_FPGA	32	12	possibly	2nd proposed method	0.012
state_space32	32	32	possibly	2nd proposed method	0.035
state_space48	48	48	possibly	2nd proposed method	0.100
cn_crr7	56	15	possibly	2nd proposed method	0.026
cn_crr10	80	21	possibly	2nd proposed method	0.064
cn_crr15	120	31	possibly	2nd proposed method	0.193
cn_crr25	200	51	possibly	2nd proposed method	0.832

with only ones in the reduced row echelon matrix obtained as a mid-result of this algorithm indicate potential unbounded places. Otherwise, it is highly likely that the Petri net is covered, and in that case, the second proposed algorithm can be used to accomplish the coverage (confirming the boundedness).

The final findings of the extensive research on the boundedness property are presented in Table **5.4** as a toolchain that combines the three novel methods examined above. The third proposed algorithm operating in polynomial time performs a preliminary boundedness analysis of a Petri net. If the Petri net is uncovered, then the use of methods based on invariant cover terminates with information about unresolved boundedness. Then, we should apply the first proposed method based on searching the reachability graph. In the case of uncovered Petri nets, it is probable that the net has unbounded places, hence it will be unnecessary to build the entire graph. In the case of a potentially covered Petri net, it is optimal to use the second novel algorithm based on the cover of invariants. As the results show, it was possible

Table 5.4 Sample results of experiments, combined proposed methods.

Name	|P|	|T |		Toolchain Bounded Applied Methods	Runtime [ms]
coloured	4	3	true	$3^{rd} + 2^{nd}$ prop. method	0.091
cortadella1	9	6	true	$3^{rd} + 2^{nd}$ prop. method	0.127
board_game	13	13	false	$3^{rd} + 1^{st}$ prop. method	15.292
lnet_p6n2	14	16	false	$3^{rd} + 1^{st}$ prop. method	49.145
dining_philosophers	15	10	false	$3^{rd} + 1^{st}$ prop. method	45.730
brenner1	16	12	true	$3^{rd} + 2^{nd}$ prop. method	0.529
state_space16	16	16	true	$3^{rd} + 2^{nd}$ prop. method	2.175
automation3	17	16	true	$3^{rd} + 1^{st}$ prop. method	78.445
balduzzi1	18	16	true	$3^{rd} + 1^{st}$ prop. method	30.214
adam1	24	12	true	$3^{rd} + 2^{nd}$ prop. method	1.892
HAN	24	40	true	$3^{rd} + 2^{nd}$ prop. method	1.241
crossroadSM_FPGA	32	12	true	$3^{rd} + 2^{nd}$ prop. method	1.461
state_space32	32	32	true	$3^{rd} + 2^{nd}$ prop. method	2.289
state_space48	48	48	true	$3^{rd} + 2^{nd}$ prop. method	8.444
cn_crr7	56	15	true	$3^{rd} + 2^{nd}$ prop. method	3.858
cn_crr10	80	21	true	$3^{rd} + 2^{nd}$ prop. method	12.782
cn_crr15	120	31	true	$3^{rd} + 2^{nd}$ prop. method	37.924
cn_crr25	200	51	true	$3^{rd} + 2^{nd}$ prop. method	207.225

to investigate the boundedness property for all 243 Petri nets. Therefore, we confirm our proposition that it is possible to examine boundedness efficiently and effectively (within the rules we have set).

5.4 SAFENESS

The two algorithms for safeness analysis proposed in Section 4.2 were experimentally verified. The effectiveness (the proper result) and efficiency (the run-time of the method) of the proposed algorithms were evaluated as a key objective of the research. The novel methods were compared with their reference method, based on exploration of the reachability graph and computation of invariants. The fourth proposed algorithm (q.v. Algorithm 4.4) explores a reachability graph for an unsafe place. When such a place is found, the algorithm terminates with a result of a Petri net being unsafe. The Martinez-Silva (M-S) algorithm for the calculation of invariants presented in Algorithm 3.3 was encouraged to obtain a state machine cover as the fifth reference method (q.v. Algorithm 3.7). The fifth method proposed in Algorithm 4.5 is based on the computation of place invariants and focuses on efficiency. It searches for invariants that form correct state machine components and, after each subsequent obtained component, verifies whether a Petri net has already been covered by those SMCs. If the Petri net is covered by state machine components, the method ends and it is claimed that the net is safe.

The algorithms were evaluated on the same dedicated computational server as for boundedness experiments. Tables **5.5** (reachability graph-based) and Table **5.6** (invariant-based) show the results of experiments for selected benchmarks. The full report of the experiments is attached in Appendix A. The particular columns of tables contain the following values:

- Name — the name of a Petri net,
- $|P|$ — the number of places in a Petri net,
- $|T|$ — the number of transitions in a Petri net,
- safe — whether the Petri net is safe,
- runtime — the run-time of the respective methods in milliseconds,
- 4th reference method — the reachability graph-based reference algorithm,
- 5th reference method — the reference classic M-S algorithm for state machine components cover verification,
- 4th proposed method — the proposed algorithm based on termination operation when an unsafe place is found,
- 5th proposed method — the proposed algorithm based on terminating operation when SMCs coverage is obtained.

The algorithms compared are divided into two groups. The first focuses on methods based on reachability graphs. They are dedicated to the criterion of effectiveness, as the methods obtain an exact result, *i.e.*, a Petri net is safe or unsafe. The fourth proposed method makes it possible to reduce the time needed to obtain the result by simultaneously building a graph and searching for an unsafe place. Table **5.5** presents experimental research that compares the fourth reference method with the fourth proposed method. For example, in the case of the unsafe Petri net *s_net_copy_mill*, a classic solution generates a result after 41 seconds (41,407.140 ms) while the proposed method returns a result in less than 0.2 s (145.046 ms). This novel algorithm finds a key application in the initial phases of the design of systems described by Petri nets. It often happens that a designer makes a mistake in the first versions of the

Table 5.5 Sample results of safeness experiments, reachability graph-based algorithms.

Name	|P|	|T |	4th Reference Method Safe Runtime[ms]		4th Proposed Method Safe Runtime [ms]	
kovalyov92	4	6	true	20.126	true	15.358
net1	6	5	true	29.288	true	16.399
CNC_machine	7	5	true	17.480	true	17.681
miczulski1	9	8	true	24.447	true	19.300
franczok1	10	14	true	32.773	true	18.367
Elevator01	12	17	false	75.668	false	64.379
board_game	13	13	false	21.769	false	14.392
hee1	14	12	true	40.923	true	56.927
silva5	16	8	true	52.875	true	36.178
hulgaard1	19	12	true	41.042	true	27.289
ConsistentExample	29	25	false	91.635	false	45.275
zuberek1	30	22	true	281.995	true	102.722
s_net_copy_mill	32	29	false	41,407.10	false	145.046
zuberek5	43	34	true	557.506	true	146.174
cn_crr15	120	31	n/a	n/a	true	n/a

specification (*e.g.*, incorrect synchronization to places, one of the most popular mistakes made by designers), and then the designed Petri net is not safe. The proposed method makes it possible to show in an efficient way that a Petri net is unsafe. It is worth noting that in the case of a safe Petri net, the operating time of the proposal is slightly shorter, especially in the case of larger nets (*e.g.*, *hulgaard1*, *zuberek5*). This is because the reachability graph is explored during its construction, so there is no need to examine the entire graph after its construction performed by means of typical available methods including the reference method (construct the reachability graph and then read the properties of the Petri net).

The second group consists of algorithms based on linear algebra. These methods are geared towards the criterion of efficiency. The result is not an exact: safe or unsafe, but safe or uncovered by state machine components. It is known from previous explanations and statements that a Petri net is safe if it is covered by SMCs. Otherwise, solutions based on these approaches are inconclusive. However, it should be added that there are relatively few cases of safe and uncovered Petri nets. In our test library of 243 Petri nets, there are 12 such cases, so they account for less than 5% of all benchmarks.

Table 5.6 Sample results of safeness experiments, linear algebra-based algorithms.

Name	|P|	|T|	5th Reference Method Covered	Runtime[ms]	5th Proposed Method Covered	Runtime[ms]
kovalyov92	4	6	true	0.026	true	0.096
net1	6	5	true	0.120	true	0.123
CNC_machine	7	5	true	0.035	true	0.086
miczulski1	9	8	true	0.058	true	0.138
franczok1	10	14	true	0.157	true	0.203
Elevator01	12	17	false	0.090	false	0.386
Board_game	13	13	false	0.107	false	0.345
hee1	14	12	true	0.489	true	0.609
silva5	16	8	true	1.449	true	0.667
hulgaard1	19	12	true	11.882	true	1.194
ConsistentExample	29	25	false	56,220.855	false	964.609
zuberek1	30	22	true	12.602	true	3.973
s_net_copy_mill	32	29	false	0.444	false	0.642
zuberek5	43	34	n/a	n/a	true	2.726
cn_crr15	120	31	n/a	n/a	true	45.246

The fifth reference method that implements the classic Martinez-Silva algorithm is compared with the fifth proposed algorithm. The proposal computes a reduced set of place invariants and forms state machine components that cover a Petri net. Hence, it is not required to obtain all the (minimum) invariants that compose the correct SMCs, but only those necessary for coverage. As experimental research shows in Table **5.6**, the results obtained by the fifth proposed algorithm are slower in time compared to the reference method. For small Petri nets, for example *kovalyov92*, *net1*, *CNC machine*, the difference can be insignificant or even negative, as additional coverage check after each state machine component can increase the running time. However, especially for larger Petri nets, such as *ConsistentExample*, the difference is significant. Moreover, for large Petri nets, such as *zuberek5* or *cn crr15*, the algorithm is still efficient in contrast to the reference method, which does not obtain a result in the assumed time, that is, one hour.

CONCLUSION

To conclude the experimental results on the property of safeness, the fourth proposed method based on the reachability graph is dedicated to effective rejection of Petri net safeness, while the fifth proposed algorithm makes it possible to efficiently analyze large (several or more place and transitions) Petri nets covered by state machine components. The limitation of the fourth proposal is that due to the requirement of building the entire graph in the case of a safe Petri net, it may not be efficient for large size nets. On the other hand, the novel method using the linear algebra technique does not present a significant gain with uncovered Petri nets, since it will be necessary to obtain all SMCs anyway. Hence, a problem arises as to which method to use in a particular case. The third proposed method from the analysis of boundedness can be helpful here in estimating in polynomial time whether a selected Petri net can be covered by place invariants, as, by definition, an unbounded Petri net cannot be safe. If a Petri net is uncovered by p-invariants, it is probably unbounded, and hence it is also unsafe. Thus, the fourth proposed method may find its best application. Otherwise, it is highly likely that the Petri net is covered and therefore bounded. In this case, the fifth proposed method is effective, as it confirms that a Petri net is not only bounded but also safe at the same time.

REFERENCES

[1] M. Wojnakowski, M. Popławski, R. Wiśniewski, and G. Bazydło, "Hippo-CPS: verification of boundedness, safeness and liveness of petri net-based cyber-physical systems," in *Technological Innovation for Digitalization and Virtualization,* L. M. Camarinha- Matos, Ed., IFIP Advances in Information and Communication Technology, Cham: Springer International Publishing, pp. 74–82, 2022.

[2] R. Wiśniewski, G. Bazydło, M. Wojnakowski, and M. Popławski, "Hippo-CPS: A tool for verification and analysis of petri net-based cyber-physical systems," in *Application and Theory of Petri Nets and Concurrency,* L. Gomes and R. Lorenz, Eds., Cham: Springer Nature Switzerland, vol. 13929, pp. 191–204, 2023.

[3] R. Wiśniewski, G. Bazydło, L. Gomes, A. Costa, and M. Wojnakowski, "Analysis and design automation of cyber-physical system with hippo and IOPT-Tools," in *IECON 2019 - 45th Annual Conference of the IEEE Industrial Electronics Society*, vol. 1, pp. 5843– 5848, 2019. ISSN: 2577-1647.

[4] W. Knottenbelt, "PIPE v2. 5: A Petri net tool for performance modelling," in *Proceedings of the 23rd Latin American Conference on Informatics, (CLEI 2007)*, Osijek, Croatia, Faculty of Law, Josip Juraj Strossmayer University in Osijek, vol. 12, 2007.

CHAPTER 6

Effective and Efficient Analysis of Boundedness and Safeness - Case Study

Abstract: This chapter deals with a case study of a Petri net-based specification of a manufacturing system. The purpose of this chapter is to introduce one of the manufacturing systems and to illustrate the proposed analytical methods on a reallife example. This chapter also provides a synthetic summary of the research with the presentation of key innovative elements of the author's over seven-year work. The chapter comments on the proposition as well as on the contribution of this book. Moreover, it outlines possible directions for further endeavors.

Keywords: Boundedness, Case study, Effectiveness, Efficiency, Experimental verification, Manufacturing system, Petri net-based specification, Safeness.

6.1 INTRODUCTION

Petri nets have gained popularity among researchers and engineers due to the combination of their graphic form of presentation with a mathematical description of operation and the possibility of modeling concurrent control systems. Based on mathematical formalism, it is possible to clearly determine the correctness and robustness of the modeled system, while graphical projection of algorithms controlling control systems promotes the ease of the design process. In addition, it is possible to automatically generate codes ready for implementation in microprocessor or FPGA devices.

The gap in effective and efficient methods o f analyzing boundedness and safeness, *i.e.*, two key properties in modeling systems specified by Petri nets, constituted a chief motivation to undertake the research described in this book. The analysis of the properties is not a trivial issue as it involves exponential computational complexity in the general case. Therefore, the aim of this work was to propose novel methods of verification, dedicated separately to the efficiency or effectiveness criterion. The introduced algorithms were implemented and compared to known reference methods by several comprehensive tests. A library of 243 Petri nets of various levels of complexity from the Hippo project constituted the basis for the

Marcin Wojnakowski

test modules applied. The full results of the experiments performed can be found in Appendix A. The novel algorithms implemented have become part of the Hippo project developed at the University of Zielona Góra.

6.2 CASE STUDY EXAMPLE

The flat bar manufacturing process consists of nine points as described in a study [1].

1. Initially, a customer reports a demand for a product with specified parameters.
2. The documentation provided by the customer is studied. Designers assess the feasibility of the order. At this stage, the documentation is accepted or changes are made after consulting with the customer.
3. The approved documentation is delivered to the implementation team and material is ordered in order to manufacture a flat bar of a length specified by the customer within the range of 3 to 6 meters.
4. The material received from the warehouse that does not meet the requirements is adapted to the specifications. The processed material is transferred to a band saw station.
5. The operator cuts the material accordingly. Production waste is stored in postproduction waste storage.
6. The semifinished product is transported for milling. The scrap material formed during the grinding process is stored in a special place.
7. The quality of the resulting product is controlled. The produced product that does not meet the requirements is disposed of as waste. The product that requires corrections is subject to additional processing. The produced product compliant with the quality parameters is transported to the pallet packing station.
8. The products are packed on pallets and other production activities are conducted.
9. Finally, the ready pallets are handed over to the customer.

With reference to the informal specification, a model of the described manufacturing system was prepared on a Petri net. A Petri net used for the specification is required to be safe thus also bounded. In addition, the net features input and output ports to communicate with the environment. Strictly speaking, it is an interpreted net. The analyzed system may be found in a production company that belongs to a group of small and medium-sized enterprises of western Poland. They specialize in manufacturing and providing services to other industrial companies. The business operates in the field of processing large-format materials by means of laser and plotter technology. The Petri net-based specification of this system is shown in Fig. (**6.1**). This Petri net consists of 24 places and 18 transitions along with four input signals and twenty output signals. Active input signals are assigned to the transitions under some conditions and are described in Table **6.2**. On the other hand, the output signals that activate certain actions are related to places and are presented in Table **6.1**. A transition is enabled as defined by Def. 2.20, including the fulfillment of the logical input conditions assigned to it. The output signals connected to a given place are activated when the token is in that place. For this reason, the net must be safe.

6.3 BOUNDEDNESS AND SAFENESS ANALYSIS

Below, a Petri net specifying the introduced manufacturing system by means of proposed algorithms is analyzed. It should be noted at the beginning that the analysis of boundedness and safeness by means of popular CAD tools, such as PIPE and IOPT-Tools, resulted in failure due to the problems described in this book. Let us begin with the analysis of boundedness, which will be followed by the analysis of safeness.

At the beginning, we used the 3rd proposed method (q.v. Algorithm 4.3), which revealed which algorithm to choose for optimal verification. This algorithm informs in polynomial time that a Petri net is uncovered by place invariants, so there are probably unbounded places in it. Hence, the 1st proposed method (q.v. Algorithm 4.1) on the reachability graph is selected. This algorithm interrupts the construction of the graph at the first unbounded place found, *i.e.*, p_{12}. This way, we know that the Petri net is unbounded, and therefore also unsafe. Places p_{12} and p_{16} (marked in red in Fig. (**6.1**) do not have output transitions. This is one of the frequent mistakes made by designers. The model must therefore be improved. We add output transition t_{19} of p_{12} and p_{16} places, along with the transition output place p_{25} (marked in green), to the corrected specification, as shown in Fig. (**6.2**). They are responsible for waste storage and are synchronized with the main production process in transition t_{13}.

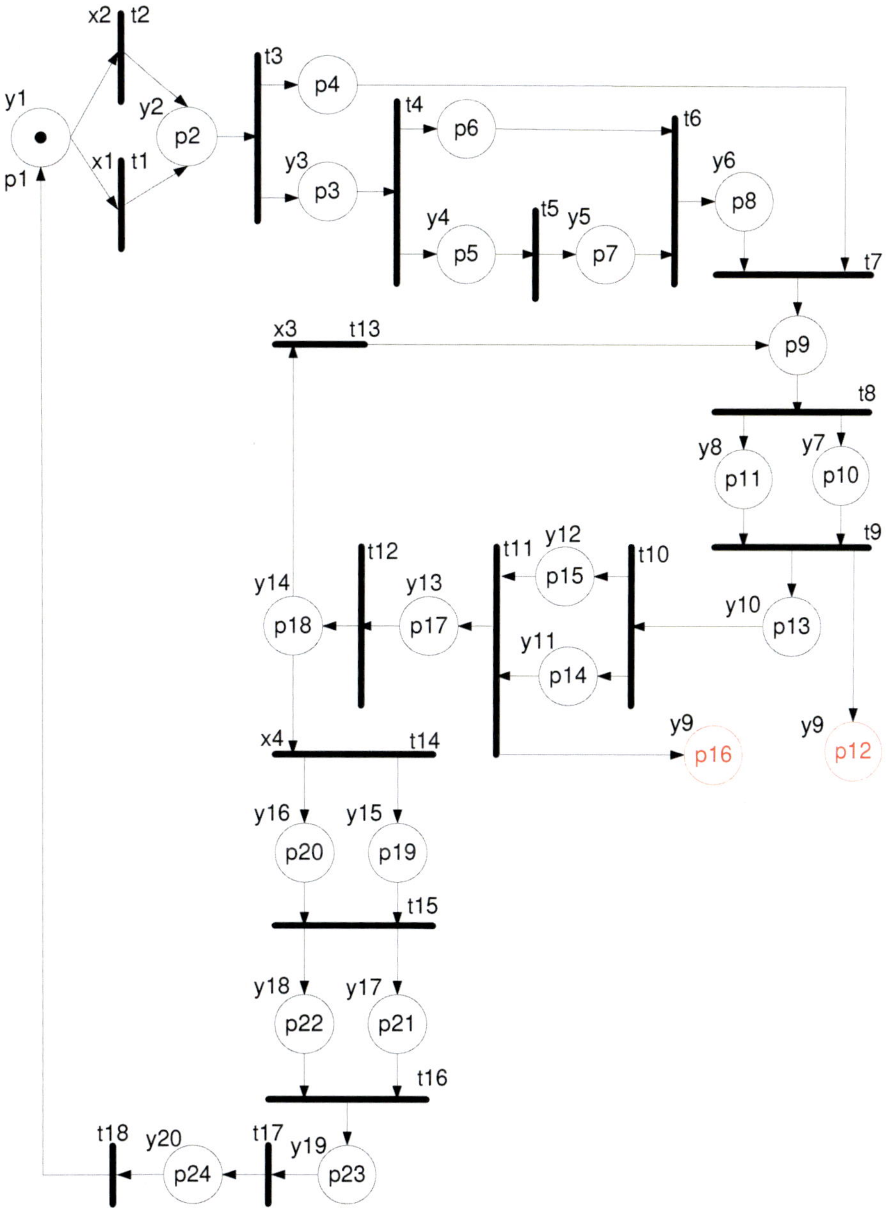

Fig. (6.1). The initial specification of a real-life manufacturing system.

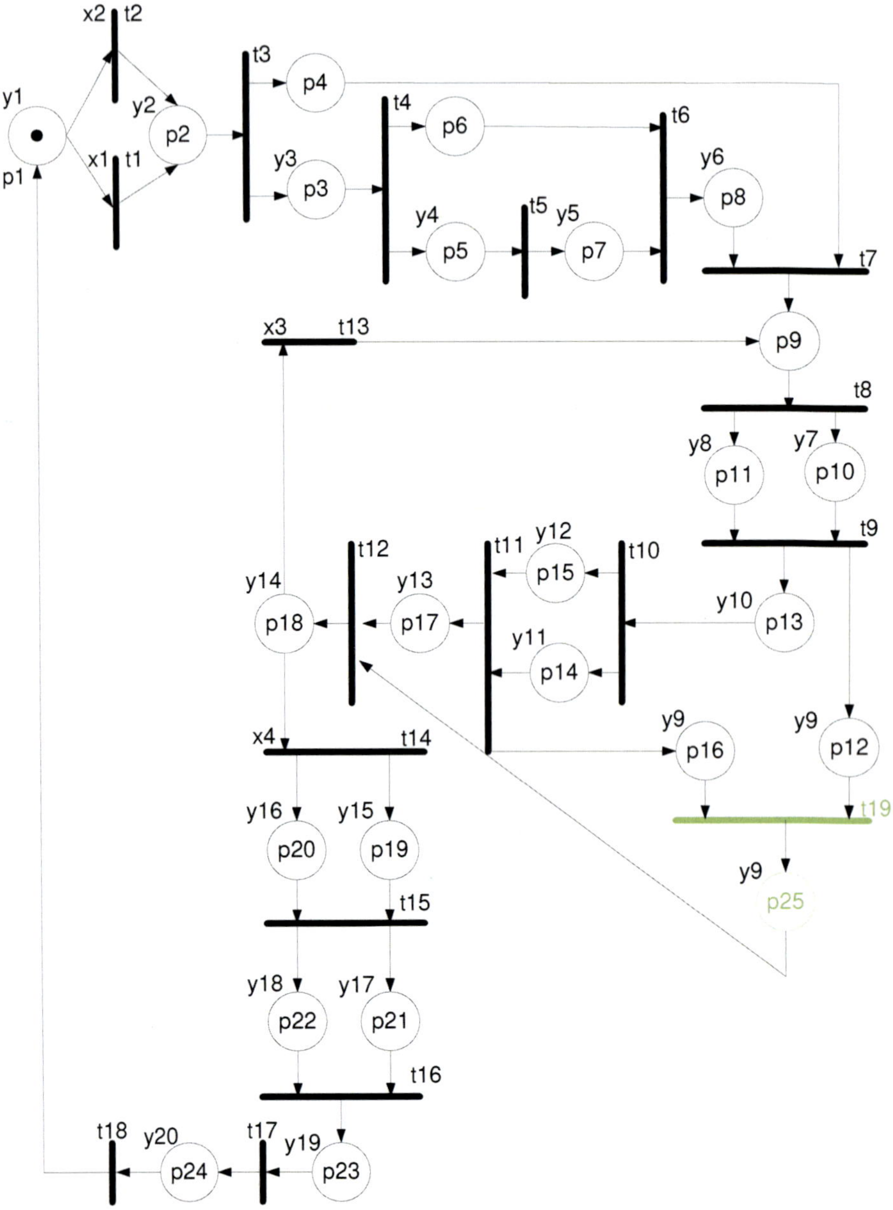

Fig. (6.2). A real-life manufacturing system specified by a Petri net.

Table 6.1 Description of output signals.

Output Signal Assigned to Task		
y_1	p_1	Request a product.
y_2	p_2	Submit the material order to the warehouse.
y_3	p_3	Send the material from the warehouse.
y_4	p_5	Take material not meeting the requirements.
y_5	p_7	Adjust material to the demanded length.
y_6	p_8	Transport the prepared material to the saw station.
y_7	p_{10}	Bring the material to the operator.
y_8	p_{11}	Cut the material accordingly.
y_9	p_{12}	Put waste to the post-production waste storage.
y_{10}	p_{13}	Transport semi-finished product for milling.
y_{11}	p_{14}	Take the operator to the station.
y_{12}	p_{15}	Process material milling.
y_9	p_{16}	Put waste to the post-production waste storage.
y_{13}	p_{17}	Transfer to the quality check control.
y_{14}	p_{18}	Verify the quality.
y_{15}	p_{19}	Move to the packing station.
y_{16}	p_{20}	Make other post-production actions.
y_{17}	p_{21}	Pack the ready product.
y_{18}	p_{22}	Put into the warehouse.
y_{19}	p_{23}	End the manufacturing procedure.
y_{20}	p_{24}	Hand the finished product down to the customer.

Table 6.2 Description of input signals.

Input Signal Assigned to Condition		
x_1	t_1	Documentation is accepted.
x_2	t_2	Changes in documentation are made.
x_3	t_{13}	Quality requirements are not met.
x_4	t_{14}	Quality requirements are met.

Subsequently, the new Petri net is reanalyzed by means of the 3rd proposed method. The algorithm finishes its operation with a result of potential coverage of the Petri net by place invariants. Then, we use the second proposed algorithm (q.v. Algorithm 4.2), which quickly finds t he Petri net covered by p-invariants and confirms the boundedness of the net. If we know that the Petri net is covered by place invariants, then the 5th proposed method (q.v. Algorithm 4.5) is applied. It checks whether the obtained invariants form correct SM-components and whether they cover the Petri net. Also, here, we obtain an affirmative answer. The analysis is terminated, and the Petri net is safe (1-bounded). The analysis of the model is possible due to the usage of proposed methods in contrast to popular solutions, *e.g.*, PIPE, used for instance, by designers of such manufacturing systems [1].

6.4 CONFIRMATION OF THE EFFECTIVE AND EFFICIENT ANALYSIS

Both the successful implementation of the selected research goals and the presentation of the related tasks confirm the correctness of the proposition. The proposed methods based on reachability graphs are dedicated to effectiveness, while the algorithms based on the invariants cover are dedicated to efficiency. Moreover, another method based on a reduced row echelon form matrix makes it possible to estimate which algorithm to apply in a particular case of a Petri net. Combining it, the proposed method using Gauss-Jordan elimination was applied, along with novel methods dedicated to efficiency and effectiveness. This created a toolchain that allowed obtaining the desired result in the assumed time of less than one hour. Hence, it was possible to analyze all 243 benchmarks from the Hippo library effectively and efficiently.

6.5 CONTRIBUTION SUMMARY

The accomplishment of the book concluded with the development, implementation, and thorough testing of a total of five novel algorithms. An overview of research results presented in this book have been initially published by the author [1-16]. The main contribution of this work can be summarized as follows:

- development and implementation of a novel algorithm for boundedness analysis of a Petri net-based specification based on a reachability graph;
- development and implementation of a novel algorithm for boundedness analysis of a Petri net-based specification based on place invariants cover (the proposed algorithm is a significant improvement of existing solutions);

- development and implementation of a novel algorithm for boundedness analysis of a Petri net-based specification based on matrix transformations and a reduced row echelon form;
- development and implementation of a toolchain combining three novel algorithms for effective and efficient boundedness analysis;
- development and implementation of a novel algorithm for safeness analysis of a Petri net-based specification based on a reachability graph;
- development and implementation of a novel algorithm for safeness analysis of a Petri net-based specification based on state machine cover (the proposed algorithm is a significant improvement of existing solutions);
- verification of effectiveness and efficiency of the proposed methods compared to reference methods;
- discussion on conclusions of the experimental research;
- participation in the development of the Hippo system, support in the process of its conception, and analysis of concurrent control systems specified by Petri nets.

CONCLUSION

Although the proposed methods are becoming a comprehensive solution and can be used in practical applications, there are still possible further research directions, as the discussed problems, such as the construction of state space or the acquisition of all minimal invariants, are characterized by exponential computational complexity in the general case. In addition to the search for further improvements in the introduced algorithms, plans may include:

- the relationship between the lack of place invariants cover and the property of boundedness for the specification of concurrent control systems;
- broader application of the novel algorithm based on a matrix in a reduced row echelon form;
- analysis of other properties of Petri nets, such as liveness;
- development of tools supporting the design of concurrent control systems described by Petri nets.

REFERENCES

[1] R. Wiśniewski, J. Patalas-Maliszewska, M. Wojnakowski, and M. Topczak, "Interpreted petri nets in modelling and analysis of physical resilient manufacturing systems," in *2022 IEEE International Conference on Systems, Man, and Cybernetics (SMC)*, pp. 1096–1102, 2022. ISSN: 2577-1655.

[2] M. Wojnakowski, R. Wiśniewski, and M. Popławski, "A Polynomial-time algorithm for detecting potentially unbounded places in a petri net-based concurrent system," in *Euro-Par 2023: Parallel Processing Workshops,* D. Zeinalipour, D. Blanco Heras, G. Pallis, H. Herodotou, D. Trihinas, D. Balouek, P. Diehl, T. Cojean, K. Fürlinger, M. H. Kirkeby, M. Nardelli, and P. Di Sanzo, Eds., Cham: Springer Nature Switzerland, pp. 277–283, 2024.

[3] M. Wojnakowski, R. Wiśniewski, G. Bazydło, and M. Popławski, "Analysis of safeness in a Petri net-based specification of the control part of cyber-physical systems," *AMCS*, vol. 31, pp. 647–657, 2021.

[4] M. Wojnakowski, M. Popławski, R. Wiśniewski, and G. Bazydło, "Hippo-CPS: verification of boundedness, safeness and liveness of petri net-based cyber-physical systems," in *Technological Innovation for Digitalization and Virtualization,* L. M. Camarinha-Matos, Ed., IFIP Advances in Information and Communication Technology, Cham: Springer International Publishing, pp. 74–82, 2022.

[5] M. Wojnakowski, M. Popławski, R. Wiśniewski, and G. Bazydło, "Safeness analysis of petri net-based cyber-physical systems based on the linear algebra and parallel reductions," in *AIP Conference Proceedings*, (Heraklion, Greece), 2021.

[6] M. Wojnakowski and R. Wiśniewski, "Verification of the boundedness property in a petri net-based specification of the control part of cyber-physical systems," in *Technological Innovation for Applied AI Systems,* L. M. Camarinha-Matos, P. Ferreira, and G. Brito, Eds., IFIP Advances in Information and Communication Technology, Cham: Springer International Publishing, pp. 83–91, 2021.

[7] M. Wojnakowski, R. Wiśniewski, and M. Popławski, "Bounded and place invariantcovered petri nets for cyber-physical systems specification," *AIP Conference Proceedings*, vol. 3030, p. 060001, 2024.

[8] M. Wojnakowski, R. Wiśniewski, M. Popławski, and G. Bazydło, "Analysis of control part of cyber-physical systems specified by interpreted petri nets," in *2022 IEEE International Conference on Systems, Man, and Cybernetics (SMC)*, pp. 1090–1095, 2022. ISSN: 2577-1655.

[9] R. Wiśniewski, G. Bazydło, L. Gomes, A. Costa, and M. Wojnakowski, "Analysis and design automation of cyber-physical system with hippo and IOPT-tools," in *IECON 2019 - 45th Annual Conference of the IEEE Industrial Electronics Society*, vol. 1, pp. 5843–5848, 2019. ISSN: 2577-1647.

[10] R. Wiśniewski, M. Wojnakowski, and Z. Li, "Design and verification of petri-net-based cyber-physical systems oriented toward implementation in field-programmable gate arrays—a case study example," *Energies*, vol. 16, p. 67, Jan. 2023.

[11] J. Patalas-Maliszewska, R. Wiśniewski, M. Topczak, and M. Wojnakowski, "Design optimization of the Petri net-based production process supported by additive manufacturing technologies," *Bull. Pol. Acad. Sci. Tech. Sci.*, vol. 70, no. 2, p. e140693, 2022.

[12] R. Wiśniewski, J. Patalas-Maliszewska, M. Wojnakowski, M. Topczak, and M. Zhou, "Fast verification of petri net-based model of industrial decision-making systems: A Case Study," in *2023 IEEE International Conference on Systems, Man, and Cybernetics (SMC)*, pp. 3316–3322, 2023. ISSN: 2577-1655.

[13] R. Wiśniewski, G. Bazydło, M. Wojnakowski, and M. Popławski, "Hippo-CPS: A tool for verification and analysis of petri net-based cyber-physical systems," in *Application and Theory of Petri Nets and Concurrency,* L. Gomes and R. Lorenz, Eds, Cham: Springer Nature Switzerland, vol. 13929, pp. 191–204, 2023.

[14] R. Wiśniewski, J. Patalas-Maliszewska, M. Wojnakowski, and M. Topczak, "Modeling and analysis of a petri net-based system supporting implementation of additive manufacturing technologies," *IEEE Transactions on Automation Science and Engineering*, pp. 1–11, 2023.

[15] R. Wiśniewski, M. Wojnakowski, and L. Stefanowicz, "Safety analysis of petri nets based on the sm-cover computed with the linear algebra technique," *AIP Conference Proceedings*, vol. 2040, p. 080008, 2018.

[16] I. Grobelna, R. Wiśniewski, and M. Wojnakowski, "Specification of cyber-physical systems with the application of interpreted nets," in *IECON 2019 - 45th Annual Conference of the IEEE Industrial Electronics Society*, vol. 1, pp. 5887–5891, 2019. ISSN: 2577-1647.

APPENDIX A

Detailed Experimental Results

This appendix includes full experimental verification of properties of boundedness and safeness.

A.1 BOUNDEDNESS

Table A.1. Full results of experiments, state space-based algorithms.

Name	$\|P\|$	$\|T\|$	Reference Method Bounded Runtime [ms]		1st Proposed Method Bounded Runtime [ms]	
prio_ex	2	3	true	56.8116	true	18.2667
semaphore	3	4	true	22.6309	true	19.2944
coloured	4	3	true	25.5637	true	17.2615
traffic_light_v2	4	3	true	18.1265	true	15.3904
pn_silva_05e	4	4	false	23.7932	false	16.9574
tank_heating	4	4	true	19.8070	true	18.3392
Entrance1	4	6	true	21.8608	true	15.8056
kovalyov92	4	6	true	20.1515	true	14.5507
return_books	4	6	true	22.5549	true	16.4531
pn_silva_05c	5	3	false	23.8781	false	16.9954
PUMA_unloading	5	3	true	57.5249	true	18.7439
consumerReachability	5	4	false	22.2833	false	21.4140
pn_silva_05b	5	4	false	33.7118	false	21.9474

(Table A.1) cont.....

pn_silva_05f	5	4	true	44462.0000	true	19.1288
pn_silva_04	5	6	false	84.3863	false	21.4031
invariants_exponent_3_2	6	2	true	20.4170	true	14.3074
np3	6	3	true	61.0780	true	18.5387
gals-example	6	4	true	58.9145	true	17.6626
lnet-p1n1	6	4	true	21.2580	true	15.8184
pcncp	6	4	true	23.5670	true	18.2273
pn-silva-02	6	4	true	55.8726	true	19.4230
PUMA-loading	6	4	true	75.6695	true	24.4826
RHINO-loading	6	4	true	59.6393	true	16.7076
RHINO-unloading	6	4	true	21.8541	true	18.7518
lnet¯p1n2	6	5	true	23.0505	true	13.9244
lnet¯p6n3	6	5	true	50.4137	true	15.4346
net1	6	5	true	62.4939	true	17.8283
traffic¯lights	6	5	true	19.8527	true	14.4626
fig311¯01	6	6	true	29.7733	true	19.0608
PNwD	6	6	true	76.1078	true	33.1066
pn¯desel¯03	6	7	true	32.4344	true	17.1911
communications¯protocol	6	8	true	25.1123	true	16.3243
health¯care¯process	6	8	true	22.3302	true	14.3417
cncrr001	7	4	true	19.9029	true	16.0355
FMS¯main¯SIPN	7	4	true	23.7193	true	15.9449
agostini1	7	5	true	18.4699	true	17.2495
CNC¯machine	7	5	true	21.2913	true	17.5161
transfer	7	5	true	24.6107	true	19.3973
voorhoeve3	7	5	true	18.0765	true	14.1707
net2 _	7	6	false	73.3893	false	26.4170
pnbrexpl 05	7	6	true	23.4051	true	20.0628
two traffic lights	7	6	true	18.2165	true	16.2687
pn desel 01	7	7	true	63.6640	true	22.6454
silva1_ _	7	7	true	49.9186	true	19.6985
traffic light v1	8	3	true	16.7578	true	16.0724
agerwala1	8	4	true	20.2320	true	16.3630
gaubert2	8	4	true	22.4147	true	15.9989
mixer mod2	8	5	true	24.3639	true	21.2758
pn desel 02	8	5	true	66.3045	true	22.9693
prod cons	8	5	true	87.3341	true	36.9736

(Table A.1) cont.....

sub-task_of_PLT_and_PMN˜	8	5	true	18.9115	true	15.4892
bridge	8	6	true	34.7380	true	21.3181
cp2	8	6	true	33.6237	true	18.8141
Exe5_split	8	6	true	61.8587	true	17.6638
img_280	8	6	true	31.9077	true	20.0653
lab5	8	6	true	26.3049	true	18.8704
TP5-I	8	6	true	19.2282	true	17.7237
bit-protocol	8	7	false	53.0126	false	27.6159
net4	8	7	true	25.4435	true	21.7310
silva14	8	7	true	61.6688	true	21.8435
elevator-2	8	8	true	26.6660	true	16.9500
girault8	8	8	true	91.0943	true	28.9878
net3	8	8	true	65.6040	true	26.5484
hard¯case	9	3	true	20.7165	true	13.7562
invariants¯exponent¯3¯3	9	3	true	21.6780	true	13.9898
bridge¯semaphore	9	6	true	27.2764	true	22.8517
cortadella1	9	6	true	29.8670	true	17.4769
lnet¯p1n4	9	6	true	31.9204	true	17.6487
multi¯robot	9	6	true	36.1574	true	24.6878
oneWayTransmissionSystem	9	6	true	81.1547	true	25.1693
semaphore2	9	6	true	63.7779	true	18.6128
simple¯production¯system	9	6	true	73.2633	true	24.9712
lnet p2n3	9	7	true	22.2145	true	16.9937
mutual exclusion	9	7	true	69.1031	true	21.5975
miczulski1	9	8	true	67.0112	true	20.4400
pnbrexpl 03	9	8	true	28.3472	true	20.9835
reactor small	9	8	true	63.2469	true	20.8127
pn silva 01	9	9	true	23.4675	true	20.5181
medeiros1	9	13	true	70.0372	true	18.9670
vanDerAalst6	9	13	true	23.0026	true	17.6228
np5	10	5	true	25.4073	true	19.3974
ConsistentExampleMessageView	10	6	true	65.9098	true	31.7462
pn silva 05d	10	6	true	62.7914	true	19.0724
exOR	10	7	true	21.0055	true	14.3336
girault4	10	7	true	41.7014	true	28.7987
speedway	10	7	true	25.8559	true	18.4950
barkaoui2	10	8	true	21.4339	true	16.9570

(Table A.1) cont.....

credit_procedure	10	8	true	87.0372	true	21.4479
lnet_p5n3	10	8	true	75.7265	true	22.5288
medeiros2	10	8	true	67.8827	true	31.7088
pascal_reach	10	8	true	68.6434	true	24.2104
silva8	10	8	true	22.0701	true	18.8941
jeng2	10	9	true	23.2713	true	17.0552
chrzastowski1	10	10	true	31.8218	true	27.5416
sivaraman1	10	10	true	70.8227	true	24.9032
franczok1	10	14	true	31.2875	true	18.2740
cncrr002	11	7	true	55.8600	true	46.5294
dataflow-computation	11	7	true	34.0984	true	20.9507
girault7	11	7	true	60.8647	true	18.1181
parallel managing automatic	11	7	false	91.7534	false	25.7083
campos1	11	8	true	37.3537	true	32.7718
julvez 1	11	8	true	130.7313	true	49.3727
lnet p5n1	11	9	true	92.4287	true	27.4539
pnbrexpl 06	11	9	true	70.7797	true	22.2867
esparza1	11	10	true	96.4341	true	36.0863
fernandez2	11	10	true	25.1146	true	18.2166
rejers2	11	10	true	71.2806	true	26.3428
bause1	11	11	true	30.3726	true	23.5162
oil separator cover s net v3	11	11	true	63.2569	true	23.0211
pnbrexpl 07	11	12	true	93.1761	true	48.4954
barkaoui1	11	13	true	33.1407	true	26.2544
li3	11	14	true	34.6593	true	21.8737
invariants exponent 3 4	12	4	true	25.3045	true	16.7715
girault1	12	6	true	29.1027	true	24.9127
real-life system smarthome	12	7	true	43.6172	true	28.1617
3carros	12	8	true	34.2907	true	31.4738
four_philosophers	12	8	true	78.9335	true	22.6871
lnet p10n1	12	9	true	115.2752	true	48.9349
machine shop	12	9	true	7184.0000	true	6.1159
campos2	12	10	true	30.1517	true	25.1777
julvez 2	12	11	true	76.8923	true	18.8027
lnet p6n1	12	11	true	85.7223	true	29.8793
bouillard1	12	12	true	47.7573	true	46.0473
li2	12	12	true	40.4389	true	23.8308

(Table A.1) cont.....

maruster3	12	13	true	66.5455	true	24.4633
ping1	12	14	true	80.2771	true	33.5221
vanDerAalst1	12	14	true	38.5573	true	28.9366
weijters1	12	14	true	36.1803	true	28.8203
Elevator01	12	17	true	115.6899	true	54.2002
lnet-p2n2	13	7	true	61.3678	true	18.5705
reiseg1	13	9	true	92.1241	true	31.0416
ExampleStateView	13	10	true	109.9310	true	48.3871
frame-manufact	13	10	true	88.5783	true	27.0384
holloway1	13	10	true	98.4501	true	40.5108
pnbrexpl-02	13	10	true	90.7901	true	35.7532
fernandez1	13	11	true	51.7186	true	18.3868
hee2	13	11	true	63.7722	true	20.3965
lnet¯p5n2	13	11	true	91.5835	true	35.2337
pnbrexpl¯08	13	11	true	76.6387	true	31.1327
pn¯campos¯silva2	13	12	true	126.0241	true	57.0109
pnbrexpl¯15	13	12	true	102.2357	true	45.4118
silva4v2	13	12	true	111.8692	true	54.4255
voorhoeve1	13	12	true	47.9559	true	44.0637
board¯game	13	13	false	21.4183	false	15.2853
hack1	13	13	true	150.5684	true	53.8082
pn campos silva7	13	13	true	91.1612	true	41.3765
pnbrexpl 04	13	13	true	97.9136	true	33.6487
silva7	13	13	true	82.0985	true	40.8650
maruster1	13	14	true	97.8088	true	32.3814
esparza3	13	15	true	110.6849	true	45.0946
girault3	14	9	true	67.2275	true	26.2178
philosophers_2_	14	10	true	97.1781	true	37.8175
philosophers 2 rev 2	14	10	true	95.8451	true	38.9143
pn silva 03	14	10	true	115.7270	true	52.3490
rejers1	14	11	true	101.2150	true	52.1882
hee1	14	12	true	81.6886	true	29.1998
voorhoeve2	14	12	true	55.5967	true	43.3879
pnbrexpl 12	14	13	true	95.2474	true	49.2921
vanDerAalst3	14	13	true	37.9714	true	29.2048
vanDerAalst7	14	13	true	55.7063	true	48.2878
jeng1	14	16	true	82.5513	true	32.6253

(Table A.1) cont.....

lnet_p6n2	14	16	false	1316.9184	false	49.1367
lnet_p9n1	14	16	true	119.1472	true	99.9595
desel1	15	9	true	378.7918	true	108.0438
dining_philosophers	15	10	false	124.8357	false	45.7244
campos3	15	11	true	60.7875	true	52.4055
oil_separator_cover_s_net_v2	15	11	true	64.4652	true	23.3883
IEC	15	12	true	44.8522	true	30.2662
esparza2	15	13	true	103.4161	true	45.9300
lasire1	15	13	true	34.1462	true	27.5741
lnet-p8n4	15	16	true	112.9582	true	63.7016
dongen1	15	17	true	43.5313	true	27.6424
2pusher	15	18	false	104.1100	false	68.5973
gaubert1	16	8	true	32.5419	true	22.1618
silva5	16	8	true	94.6754	true	38.8790
girault2	16	10	true	97.8294	true	31.0547
brenner1	16	12	true	51507.5962	true	49074.9953
beverage	16	13	true	54.3064	true	42.5259
beverage_production_part2	16	13	true	55.4875	true	41.2558
maruster4	16	13	true	165.2804	true	59.8852
mixer_one_cup	16	13	true	95.4623	true	54.5898
pnbrexpl_09	16	13	true	111.0386	true	53.8896
chiola1	16	15	true	32.3218	true	24.3016
state space16	16	16	true	3208.8901	true	1802.1074
handling of incoming order	17	14	true	86.4162	true	28.1294
eshuis1	17	15	true	76.8377	true	32.9730
automation3	17	16	true	66.7310	true	78.4346
s net frame manufact quality~	17	16	true	38.4612	true	31.0045
li1	17	17	true	113.4483	true	64.7912
pn campos silva6	18	11	true	2692509.0000	true	8.4306
barkaoui3	18	13	true	68.4313	true	50.8150
balduzzi1	18	16	true	31.6941	true	30.2047
vanDerAalst4	18	16	true	89.3019	true	79.2355
kavi1	18	20	true	352.2237	true	118.4768
PWM patterns	19	11	true	89.2849	true	32.5728
hulgaard1	19	12	true	81.2282	true	26.0502
mixer	19	15	true	114.7682	true	56.8489
s net frame manufact quality~	19	18	true	52.6186	true	43.4010

(Table A.1) cont.....

maruster2	20	14	true	116.9708	true	60.1574
holliday1	20	15	true	519.5248	true	176.0140
philosophers_5	20	15	true	169.5814	true	81.3015
beverage_production	20	16	true	59.5669	true	44.9109
miling_machine	20	16	true	186.6802	true	68.1431
milling	20	16	true	203.7742	true	65.6276
mixer-mod1	20	16	true	111.4485	true	53.7371
dingle1	20	20	true	83.8090	true	29.0316
basile1	21	17	true	107.5966	true	86.9539
lnet-p8n3	21	17	true	207.1881	true	73.0312
fernandez3	21	20	true	90.4209	true	29.5087
pn-fernandez3	21	20	true	95.5674	true	36.9348
heiner1	22	20	true	283.3456	true	95.5728
chrzastowski2	22	21	true	46.9244	true	44.2883
vanDerAalst5	23	23	true	78.8984	true	55.8145
adam1	24	12	true	57.1502	true	55.1834
xie1	24	26	true	116.9825	true	107.4893
HAN	24	40	true	100185.7668	true	50454.5466
oil_separator_cover_alfa_net	27	21	true	98.8265	true	51.1180
lnet_p8n2	28	17	true	209.8969	true	80.9932
zuberek3	29	21	true	201.6953	true	179.4626
alfa_net_copy_milling__quality	29	25	false	67.7053	false	55.6947
ConsistentExample	29	25	true	180.9862	true	80.8641
oil separator cover s net	29	25	true	87.3408	true	56.2373
zuberek4	30	21	true	225.1568	true	194.4964
zuberek1	30	22	true	121.8420	true	107.7838
well_built_alfa_net˜ subprocess	30	25	true	141.6397	true	139.1226
alfa_net copy milling subprocess	30	26	true	48.6994	true	42.9979
s net copy milling˜ subprocess	31	28	true	154.7889	true	176.1151
crossroadSM_FPGA	32	12	true	79.0077	true	30.4632
alfa net copy milling˜ synchr	32	27	true	192.1757	true	207.2063
s net_copy milling˜ quality	32	29	false	23633.6107	false	142.5313
state space32	32	32	n/a	n/a	n/a	n/a
vanDerAalst8	34	27	true	139.9571	true	120.1210
lnet p4n1	37	41	true	105.9426	true	39.7104
zuberek5	41	31	true	175.7267	true	154.3409
lnet p7n1	41	32	true	269648.5325	true	129680.3014

(Table A.1) cont.....

state_space48	48	48	n/a	n/a	n/a	n/a
PWM extended	49	31	true	1110.2623	true	293.2315
lnet_p8n1	51	40	true	42494.7059	true	22073.1653
cn_crr7	56	15	true	191885.1152	true	95667.5924
cn_crr10	80	21	n/a	n/a	n/a	n/a
cn_crr15	120	31	n/a	n/a	n/a	n/a
cn_crr25	200	51	n/a	n/a	n/a	n/a

Table A.2. Full results of experiments, linear algebra-based algorithms.

Name	|P|	|T|	Reference Method Covered Runtime[ms]		2nd Proposed Method Covered Runtime[ms]	
prio ex	2	3	true	0.022	true	0.087
semāphore	3	4	true	0.053	true	0.084
coloured	4	3	true	0.045	true	0.088
traffic light v2	4	3	true	0.031	true	0.087
pn silv̄a 05ē	4	4	false	0.061	false	0.603
tank_heating	4	4	true	0.063	true	0.079
Entrance1	4	6	true	0.022	true	0.114
kovalyov92	4	6	true	0.024	true	0.086
return_books	4	6	true	0.021	true	0.092
pn_silva_05c	5	3	false	0.042	false	0.093
PUMA_unloading	5	3	true	0.080	true	0.082
consumerReachability	5	4	false	0.076	false	0.115
pn_silva_05b	5	4	false	0.061	false	0.099
pn_silva_05f	5	4	true	0.026	true	0.101
pn_silva_04	5	6	false	0.023	false	0.336
invariants_exponent_3_2	6	2	true	0.041	true	0.096
np3	6	3	true	0.074	true	0.119
gals-example	6	4	true	0.028	true	0.119
lnet-p1n1	6	4	true	0.034	true	0.145
pcncp	6	4	true	0.070	true	0.116
pn-silva-02	6	4	true	0.043	true	0.089
PUMA-loading	6	4	true	0.055	true	0.087
RHINO-loading	6	4	true	0.139	true	0.085

(Table A.2) cont.....

RHINO_unloading	6	4	true	0.070	true	0.091
lnet_p1n2	6	5	true	0.145	true	0.108
lnet_p6n3	6	5	true	0.049	true	0.103
net1	6	5	true	0.115	true	0.109
traffic_lights	6	5	true	0.084	true	0.134
fig311_01	6	6	true	0.030	true	0.141
PNwD	6	6	true	0.034	true	0.149
pn-desel-03	6	7	true	0.033	true	0.125
communications-protocol	6	8	true	0.027	true	0.126
health-care-process	6	8	true	0.059	true	0.176
cncrr001	7	4	true	0.040	true	0.187
FMS-main-SIPN	7	4	true	0.071	true	0.177
agostini1	7	5	true	0.042	true	0.121
CNC-machine	7	5	true	0.033	true	0.092
transfer	7	5	false	0.036	false	0.135
voorhoeve3	7	5	true	0.174	true	0.101
net2	7	6	false	0.052	false	0.102
pnbrexpl_05	7	6	true	0.051	true	0.305
two_traffic_lights	7	6	true	0.062	true	0.126
pn_desel_01	7	7	true	0.038	true	0.096
silva1	7	7	true	0.076	true	0.103
traffic_light_v1	8	3	true	0.064	true	2.752
agerwala1	8	4	true	0.041	true	0.167
gaubert2	8	4	true	0.058	true	0.168
mixer_mod2	8	5	true	0.038	true	7.387
pn_desel_02	8	5	true	0.041	true	0.132
prod_cons	8	5	true	0.040	true	0.139
sub-task of PLT and PMN	8	5	true	0.037	true	0.143
bridge	8	6	true	0.059	true	0.100
cp2	8	6	false	0.073	false	0.139
Exe5 split	8	6	true	0.065	true	0.099
img 280	8	6	true	0.042	true	0.117
lab5_	8	6	true	0.042	true	0.142
TP5 I	8	6	true	0.047	true	0.103
bit protocol	8	7	false	0.040	false	0.240
net4	8	7	true	0.120	true	0.157
silva14	8	7	true	0.051	true	0.494

(Table A.2) cont.....

elevator_2	8	8	true	0.032	true	0.105
girault8	8	8	true	0.045	true	0.158
net3	8	8	true	0.040	true	0.160
hard_case	9	3	true	0.214	true	0.125
invariants_exponent_3_3	9	3	true	0.104	true	0.131
bridge-semaphore	9	6	true	0.069	true	0.212
cortadella1	9	6	true	0.065	true	0.124
lnet-p1n4	9	6	true	0.071	true	0.142
multi-robot	9	6	true	0.052	true	0.171
oneWayTransmissionSystem	9	6	true	0.146	true	0.174
semaphore2	9	6	true	0.084	true	0.138
simple-production-system	9	6	true	0.092	true	0.207
lnet-p2n3	9	7	true	0.153	true	0.130
mutual-exclusion	9	7	true	0.044	true	0.235
miczulski1	9	8	true	0.055	true	0.132
pnbrexpl-03	9	8	true	0.110	true	0.276
reactor-small	9	8	true	0.134	true	0.132
pn-silva-01	9	9	true	0.062	true	0.150
medeiros1	9	13	true	0.049	true	0.192
vanDerAalst6	9	13	true	0.058	true	0.516
np5	10	5	true	0.086	true	0.220
ConsistentExampleMessageView	10	6	true	0.132	true	0.221
pn silva 05d	10	6	false	0.091	false	0.207
exOR	10	7	false	0.066	false	0.260
girault4	10	7	true	0.053	true	0.186
speedway	10	7	true	0.506	true	0.246
barkaoui2	10	8	true	0.064	true	0.225
credit procedure	10	8	true	0.046	true	0.133
lnet p5n3	10	8	true	0.223	true	0.142
medeiros2	10	8	true	0.067	true	0.251
pascal reach	10	8	true	0.080	true	1.150
silva8	10	8	false	0.067	false	0.161
jeng2	10	9	true	0.080	true	0.164
chrzastowski1	10	10	true	0.045	true	0.242
sivaraman1	10	10	true	0.047	true	0.146
franczok1	10	14	true	0.151	true	0.193
cncrr002	11	7	true	0.075	true	0.259

(Table A.2) cont.....

dataflow_computation	11	7	true	0.066	true	0.235
girault7	11	7	true	0.107	true	0.174
parallel_managing_automatic_devices	11	7	false	0.077	false	0.209
campos1	11	8	true	2.458	true	0.335
julvez_1	11	8	true	0.066	true	0.257
lnet_p5n1	11	9	true	0.047	true	0.157
pnbrexpl_06	11	9	true	0.090	true	0.164
esparza1	11	10	true	0.055	true	0.324
fernandez2	11	10	true	0.056	true	0.207
rejers2	11	10	true	0.083	true	0.137
bause1	11	11	false	0.055	false	0.283
oil¯separator¯cover¯s¯net¯v3	11	11	true	0.072	true	0.221
pnbrexpl¯07	11	12	true	0.098	true	0.190
barkaoui1	11	13	true	0.063	true	0.243
li3	11	14	true	0.091	true	0.234
invariants¯exponent¯3¯4	12	4	true	0.635	true	0.137
girault1	12	6	true	0.084	true	0.350
real-life¯system¯smarthome˜	12	7	true	0.081	true	0.194
3carros	12	8	true	0.090	true	0.116
four¯philosophers	12	8	true	0.092	true	0.319
lnet¯p10n1	12	9	true	0.073	true	0.129
machine¯shop	12	9	false	0.168	false	0.315
campos2	12	10	true	0.058	true	0.318
julvez 2	12	11	true	0.082	true	0.560
lnet p6n1	12	11	true	0.074	true	0.160
bouillard1	12	12	true	0.077	true	0.246
li2	12	12	true	0.067	true	0.136
maruster3	12	13	true	0.153	true	0.164
ping1	12	14	true	0.085	true	0.259
vanDerAalst1	12	14	true	0.092	true	0.230
weijters1	12	14	true	0.171	true	0.507
Elevator01	12	17	false	0.087	false	1.108
lnet p2n2	13	7	true	0.137	true	0.289
reiseg1	13	9	true	0.098	true	0.265
ExampleStateView	13	10	false	0.070	false	0.315
frame manufact	13	10	true	0.185	true	0.204
holloway1	13	10	true	0.938	true	0.218

(Table A.2) cont.....

pnbrexpl_02	13	10	true	0.083	true	0.135
fernandez1	13	11	true	0.155	true	0.297
hee2	13	11	true	0.159	true	0.470
lnet_p5n2	13	11	true	0.219	true	0.172
pnbrexpl_08	13	11	true	0.096	true	0.174
pn_campos_silva2	13	12	true	0.078	true	4.127
pnbrexpl_15	13	12	true	0.274	true	0.247
silva4v2	13	12	true	0.073	true	0.421
voorhoeve1	13	12	true	0.211	true	0.282
board_game	13	13	false	0.104	false	0.237
hack1	13	13	true	0.100	true	0.826
pn_campos_silva7	13	13	true	0.372	true	0.316
pnbrexpl_04	13	13	true	0.131	true	0.281
silva7	13	13	true	0.116	true	0.313
maruster1	13	14	true	0.128	true	0.278
esparza3	13	15	true	0.134	true	0.385
girault3	14	9	true	0.319	true	0.946
philosophers_2	14	10	true	0.107	true	0.463
philosophers_2_rev_2	14	10	true	0.112	true	0.473
pn_silva_03	14	10	true	0.290	true	0.321
rejers1	14	11	true	0.358	true	0.134
hee1	14	12	true	0.475	true	0.567
voorhoeve2	14	12	true	0.429	true	0.321
pnbrexpl 12	14	13	true	0.090	true	0.248
vanDerAalst3	14	13	true	0.125	true	0.138
vanDerAalst7	14	13	true	0.067	true	0.566
jeng1	14	16	true	0.061	true	0.309
lnet_p6n2	14	16	false	0.085	false	0.240
lnet p9n1	14	16	true	0.201	true	0.264
desel1	15	9	true	0.061	true	0.157
dining philosophers	15	10	false	0.068	false	0.331
campos3	15	11	true	0.131	true	0.566
oil separator cover s net v2	15	11	true	0.388	true	3.051
IEC	15	12	true	0.110	true	0.213
esparza2	15	13	true	2.277	true	0.544
lasire1	15	13	true	0.117	true	0.180
lnet p8n4	15	16	true	0.081	true	0.398

(Table A.2) cont.....

dongen1	15	17	true	0.068	true	0.319
2pusher	15	18	false	0.109	false	1.688
gaubert1	16	8	true	0.260	true	0.490
silva5	16	8	true	1.443	true	0.546
girault2	16	10	true	0.170	true	0.537
brenner1	16	12	true	0.095	true	0.519
beverage	16	13	true	0.094	true	0.272
beverage-production-part2	16	13	true	0.085	true	0.226
maruster4	16	13	true	0.164	true	0.214
mixer-one-cup	16	13	true	0.078	true	10.047
pnbrexpl-09	16	13	true	0.229	true	0.259
chiola1	16	15	true	0.076	true	0.573
state-space16	16	16	true	0.091	true	2.167
handling-of-incoming-order	17	14	true	0.087	true	0.403
eshuis1	17	15	true	0.130	true	0.327
automation3	17	16	false	0.083	false	0.567
s-net-frame-manufact-quality-v1	17	16	true	0.127	true	3.829
li1	17	17	true	0.546	true	0.251
pn-campos-silva6	18	11	true	0.140	true	0.590
barkaoui3	18	13	true	0.094	true	0.777
balduzzi1	18	16	false	0.107	false	0.656
vanDerAalst4	18	16	true	0.148	true	0.343
kavi1	18	20	true	0.222	true	0.503
PWM patterns	19	11	true	0.280	true	0.313
hulgaard1	19	12	true	11.863	true	0.973
mixer	19	15	true	0.106	true	0.320
s net frame manufact quality v2	19	18	true	0.129	true	0.333
maruster2	20	14	true	0.162	true	0.563
holliday1	20	15	true	0.274	true	0.631
philosophers 5	20	15	true	0.123	true	0.555
beverage production	20	16	true	0.158	true	0.346
miling machine	20	16	true	0.137	true	0.193
milling	20	16	true	0.160	true	0.187
mixer mod1	20	16	true	0.138	true	0.371
dingle1	20	20	true	0.100	true	0.895
basile1	21	17	true	0.145	true	0.422
lnet p8n3	21	17	true	0.172	true	0.328

(Table A.2) cont.....

fernandez3	21	20	true	0.225	true	0.674
pn_fernandez3	21	20	true	0.202	true	0.927
heiner1	22	20	true	0.138	true	0.900
chrzastowski2	22	21	true	0.105	true	0.223
vanDerAalst5	23	23	true	0.174	true	0.726
adam1	24	12	true	1.805	true	1.884
xie1	24	26	true	1.285	true	0.631
HAN	24	40	true	0.102	true	1.235
oil-separator-cover-alfa-net	27	21	true	1.229	true	0.468
lnet-p8n2	28	17	true	0.394	true	4.591
zuberek3	29	21	true	0.568	true	0.680
alfa_net_copy_milling~_quality	29	25	false	0.378	false	0.571
ConsistentExample	29	25	false	56 220.831	false	1 197.142
oil_separator_cover_s_net	29	25	true	0.406	true	0.559
zuberek4	30	21	true	1.920	true	1.324
zuberek1	30	22	true	12.592	true	1.647
well_built_alfa_net_copy~_subprocess	30	25	true	0.263	true	0.537
alfa_net_copy_milling_subprocess	30	26	false	0.365	false	0.587
s_net_copy_milling~_subprocess	31	28	true	0.513	true	0.874
crossroadSM_FPGA	32	12	true	891 400.791	true	1.449
alfa_net_copy_milling~_synchr	32	27	true	0.363	true	0.644
s_net_copy_milling~_quality	32	29	false	0.438	false	0.640
state space32	32	32	true	0.264	true	2.254
vanDerAalst8	34	27	true	0.414	true	4.785
lnet p4n1	37	41	true	0.170	true	4.689
zuberek5	41	31	n/a	n/a	true	2.391
lnet p7n1	41	32	true	0.631	true	1.439
state space48	48	48	true	0.375	true	8.344
PWM extended	49	31	true	2.392	true	17.842
lnet p8n1	51	40	false	1.068	false	9.713
cn_crr7	56	15	n/a	n/a	true	3.832
cn_crr10	80	21	n/a	n/a	true	12.782
cn_crr15	120	31	n/a	n/a	true	37.731
cn crr25	200	51	n/a	n/a	true	206.393

Table A.3. Full results of experiments, the reduced row echelon form-based algorithm.

Name	|P|	|T|	3rd Proposed Method Covered	Further Analysis Runtime [ms]
prio_ex	2	3	possibly 2nd proposed method	0.003
semaphore	3	4	possibly 2nd proposed method	0.003
coloured	4	3	possibly 2nd proposed method	0.003
traffic_light_v2	4	3	possibly 2nd proposed method	0.003
pn_silva_05e	4	4	false 1st proposed method	0.003
tank_heating	4	4	possibly 2nd proposed method	0.003
Entrance1	4	6	possibly 2nd proposed method	0.004
kovalyov92	4	6	possibly 2nd proposed method	0.003
return_books	4	6	possibly 2nd proposed method	0.003
pn_silva_05c	5	3	false 1st proposed method	0.003
PUMA_unloading	5	3	possibly 2nd proposed method	0.003
consumerReachability	5	4	false 1st proposed method	0.003
pn_silva_05b	5	4	false 1st proposed method	0.003
pn_silva_05f	5	4	possibly 2nd proposed method	0.003
pn_silva_04	5	6	false 1st proposed method	0.003
invariants_exponent_3_2	6	2	possibly 2nd proposed method	0.003
np3	6	3	possibly 2nd proposed method	0.003
gals-example	6	4	possibly 2nd proposed method	0.005
lnet_p1n1	6	4	possibly 2nd proposed method	0.003
pcncp	6	4	possibly 2nd proposed method	0.003
pn_silva_02	6	4	possibly 2nd proposed method	0.004
PUMA_loading	6	4	possibly 2nd proposed method	0.003
RHINO_loading	6	4	possibly 2nd proposed method	0.003
RHINO_unloading	6	4	possibly 2nd proposed method	0.003
lnet_p1n2	6	5	possibly 2nd proposed method	0.003
lnet_p6n3	6	5	possibly 2nd proposed method	0.003
net1	6	5	possibly 2nd proposed method	0.003
traffic_lights	6	5	possibly 2nd proposed method	0.004
fig311_01	6	6	possibly 2nd proposed method	0.004
PNwD	6	6	possibly 2nd proposed method	0.003
pn_desel_03	6	7	possibly 2nd proposed method	0.004
communications_protocol	6	8	possibly 2nd proposed method	0.004

(Table A.3) cont.....

health_care_process	6	8	possibly 2nd proposed method	0.003
cncrr001	7	4	possibly 2nd proposed method	0.003
FMS_main_SIPN	7	4	possibly 2nd proposed method	0.003
agostini1	7	5	possibly 2nd proposed method	0.004
CNC_machine	7	5	possibly 2nd proposed method	0.004
transfer	7	5	possibly $1^{s}t$ proposed method	0.003
voorhoeve3	7	5	possibly 2nd proposed method	0.003
net2	7	6	false 1st proposed method	0.003
pnbrexpl-05	7	6	possibly 2nd proposed method	0.003
two-traffic-lights	7	6	possibly 2nd proposed method	0.003
pn-desel-01	7	7	possibly 2nd proposed method	0.004
silva1	7	7	possibly 2nd proposed method	0.004
traffic-light-v1	8	3	possibly 2nd proposed method	0.003
agerwala1	8	4	possibly 2nd proposed method	0.003
gaubert2	8	4	possibly 2nd proposed method	0.003
mixer-mod2	8	5	possibly 2nd proposed method	0.003
pn-desel-02	8	5	possibly 2nd proposed method	0.004
prod-cons	8	5	possibly 2nd proposed method	0.004
sub-task-of-PLT-and-PMN~	8	5	possibly 2nd proposed method	0.003
bridge	8	6	possibly 2nd proposed method	0.003
cp2	8	6	possibly 1st proposed method	0.003
Exe5-split	8	6	possibly 2nd proposed method	0.004
img-280	8	6	possibly 2nd proposed method	0.003
lab5	8	6	possibly 2nd proposed method	0.003
TP5 I	8	6	possibly 2nd proposed method	0.004
bit protocol	8	7	false 1st proposed method	0.004
net4	8	7	possibly 2nd proposed method	0.003
silva14 _	8	7	possibly 2nd proposed method	0.004
elevator 2	8	8	possibly 2nd proposed method	0.004
girault8	8	8	possibly 2nd proposed method	0.004
net3 _	8	8	possibly 2nd proposed method	0.003
hard case _ _ _	9	3	possibly 2nd proposed method	0.003
invariants exponent 3 3	9	3	possibly 2nd proposed method	0.003
bridge semaphore	9	6	possibly 2nd proposed method	0.003
cortadella1	9	6	possibly 2nd proposed method	0.003
lnet p1n4	9	6	possibly 2nd proposed method	0.004
multi robot	9	6	possibly 2nd proposed method	0.004

(Table A.3) cont.....

oneWayTransmissionSystem	9	6	possibly 2nd proposed method	0.004
semaphore2	9	6	possibly 2nd proposed method	0.004
simple_production_system	9	6	possibly 2nd proposed method	0.004
lnet_p2n3	9	7	possibly 2nd proposed method	0.004
mutual_exclusion	9	7	possibly 2nd proposed method	0.004
miczulski1	9	8	possibly 2nd proposed method	0.004
pnbrexpl-03	9	8	possibly 2nd proposed method	0.004
reactor-small	9	8	possibly 2nd proposed method	0.004
pn-silva-01	9	9	possibly 2nd proposed method	0.004
medeiros1	9	13	possibly 2nd proposed method	0.005
vanDerAalst6	9	13	possibly 2nd proposed method	0.005
np5	10	5	possibly 2nd proposed method	0.003
ConsistentExampleMessageView	10	6	possibly 2nd proposed method	0.004
pn-silva-05d	10	6	possibly 1st proposed method	0.003
exOR	10	7	possibly 1st proposed method	0.005
girault4	10	7	possibly 2nd proposed method	0.004
speedway	10	7	possibly 2nd proposed method	0.004
barkaoui2	10	8	possibly 2nd proposed method	0.004
credit_procedure	10	8	possibly 2nd proposed method	0.005
lnet_p5n3	10	8	possibly 2nd proposed method	0.004
medeiros2	10	8	possibly 2nd proposed method	0.004
pascal_reach	10	8	possibly 2nd proposed method	0.004
silva8	10	8	possibly 1st proposed method	0.004
jeng2	10	9	possibly 2nd proposed method	0.005
chrzastowski1	10	10	possibly 2nd proposed method	0.005
sivaraman1	10	10	possibly 2nd proposed method	0.005
franczok1	10	14	possibly 2nd proposed method	0.006
cncrr002_	11	7	possibly 2nd proposed method	0.004
dataflow computation	11	7	possibly 2nd proposed method	0.004
girault7_	11	7	possibly 2nd proposed method	0.004
parallel managing automatic	11	7	false 1st proposed method	0.004
campos1	11	8	possibly 2nd proposed method	0.004
julvez 1	11	8	possibly 2nd proposed method	0.004
lnet p5n1	11	9	possibly 2nd proposed method	0.005
pnbrexpl 06	11	9	possibly 2nd proposed method	0.005
esparza1	11	10	possibly 2nd proposed method	0.005
fernandez2	11	10	possibly 2nd proposed method	0.006

(Table A.3) cont.....

rejers2	11	10	possibly 2nd proposed method	0.005
bause1	11	11	possibly 1st proposed method	0.005
oil_separator_cover_s_net_v3	11	11	possibly 2nd proposed method	0.005
pnbrexpl_07	11	12	possibly 2nd proposed method	0.005
barkaoui1	11	13	possibly 2nd proposed method	0.006
li3	11	14	possibly 2nd proposed method	0.007
invariants-exponent-3-4	12	4	possibly 2nd proposed method	0.004
girault1	12	6	possibly 2nd proposed method	0.004
real-life-system-smarthome˜	12	7	possibly 2nd proposed method	0.004
3carros	12	8	possibly 2nd proposed method	0.005
four-philosophers	12	8	possibly 2nd proposed method	0.004
lnet-p10n1	12	9	possibly 2nd proposed method	0.005
machine-shop	12	9	possibly 1st proposed method	0.005
campos2	12	10	possibly 2nd proposed method	0.005
julvez-2	12	11	possibly 2nd proposed method	0.005
lnet-p6n1	12	11	possibly 2nd proposed method	0.005
bouillard1	12	12	possibly 2nd proposed method	0.006
li2	12	12	possibly 2nd proposed method	0.006
maruster3	12	13	possibly 2nd proposed method	0.006
ping1	12	14	possibly 2nd proposed method	0.007
vanDerAalst1	12	14	possibly 2nd proposed method	0.007
weijters1	12	14	possibly 2nd proposed method	0.007
Elevator01	12	17	possibly 1st proposed method	0.007
lnet p2n2	13	7	possibly 2nd proposed method	0.004
reiseg1	13	9	possibly 2nd proposed method	0.005
ExampleStateView	13	10	possibly 1st proposed method	0.005
frame manufact	13	10	possibly 2nd proposed method	0.006
holloway1	13	10	possibly 2nd proposed method	0.006
pnbrexpl 02	13	10	possibly 2nd proposed method	0.006
fernandez1	13	11	possibly 2nd proposed method	0.006
hee2	13	11	possibly 2nd proposed method	0.005
lnet p5n2	13	11	possibly 2nd proposed method	0.007
pnbrexpl 08	13	11	possibly 2nd proposed method	0.006
pn campos silva2	13	12	possibly 2nd proposed method	0.006
pnbrexpl 15	13	12	possibly 2nd proposed method	0.006
silva4v2	13	12	possibly 2nd proposed method	0.006
voorhoeve1	13	12	possibly 2nd proposed method	0.006

(Table A.3) cont.....

board_game	13	13	false	1st proposed method	0.007
hack1	13	13	possibly	2nd proposed method	0.006
pn_campos_silva7	13	13	possibly	2nd proposed method	0.007
pnbrexpl_04	13	13	possibly	2nd proposed method	0.007
silva7	13	13	possibly	2nd proposed method	0.007
maruster1	13	14	possibly	2nd proposed method	0.007
esparza3	13	15	possibly	2nd proposed method	0.008
girault3	14	9	possibly	2nd proposed method	0.005
philosophers-2-rev-2	14	10	possibly	2nd proposed method	0.006
philosophers-2	14	10	possibly	2nd proposed method	0.006
pn-silva-03	14	10	possibly	2nd proposed method	0.005
rejers1	14	11	possibly	2nd proposed method	0.006
hee1	14	12	possibly	2nd proposed method	0.006
voorhoeve2	14	12	possibly	2nd proposed method	0.006
pnbrexpl-12	14	13	possibly	2nd proposed method	0.007
vanDerAalst3	14	13	possibly	2nd proposed method	0.008
vanDerAalst7	14	13	possibly	2nd proposed method	0.007
jeng1	14	16	possibly	2nd proposed method	0.009
lnet_p6n2	14	16	false	1st proposed method	0.008
lnet_p9n1	14	16	possibly	2nd proposed method	0.008
desel1	15	9	possibly	2nd proposed method	0.012
dining_philosophers	15	10	false	1st proposed method	0.006
campos3	15	11	possibly	2nd proposed method	0.006
oil_separator_cover_s_net_v2	15	11	possibly	2nd proposed method	0.006
IEC	15	12	possibly	2nd proposed method	0.007
esparza2	15	13	possibly	2nd proposed method	0.008
lasire1	15	13	possibly	2nd proposed method	0.007
lnet_p8n4	15	16	possibly	2nd proposed method	0.009
dongen1	15	17	possibly	2nd proposed method	0.010
2pusher	15	18	false	1st proposed method	0.009
gaubert1	16	8	possibly	2nd proposed method	0.005
silva5	16	8	possibly	2nd proposed method	0.005
girault2	16	10	possibly	2nd proposed method	0.007
brenner1	16	12	possibly	2nd proposed method	0.010
beverage production part2	16	13	possibly	2nd proposed method	0.009
beverage	16	13	possibly	2nd proposed method	0.008
maruster4	16	13	possibly	2nd proposed method	0.009

(Table A.3) cont.....

mixer_one_cup	16	13	possibly 2nd proposed method	0.008
pnbrexpl_09	16	13	possibly 2nd proposed method	0.008
chiola1	16	15	possibly 2nd proposed method	0.009
state_space16	16	16	possibly 2nd proposed method	0.008
handling_of_incoming_order	17	14	possibly 2nd proposed method	0.009
eshuis1	17	15	possibly 2nd proposed method	0.010
automation3	17	16	possibly 1st proposed method	0.010
s-net-frame-manufact	17	16	possibly 2nd proposed method	0.010
li1	17	17	possibly 2nd proposed method	0.011
pn-campos-silva6	18	11	possibly 2nd proposed method	0.007
barkaoui3	18	13	possibly 2nd proposed method	0.009
balduzzi1	18	16	possibly 1st proposed method	0.009
vanDerAalst4	18	16	possibly 2nd proposed method	0.011
kavi1	18	20	possibly 2nd proposed method	0.013
PWM¯patterns	19	11	possibly 2nd proposed method	0.007
hulgaard1	19	12	possibly 2nd proposed method	0.008
mixer	19	15	possibly 2nd proposed method	0.011
s¯net¯frame¯manufact	19	18	possibly 2nd proposed method	0.013
maruster2	20	14	possibly 2nd proposed method	0.010
holliday1	20	15	possibly 2nd proposed method	0.010
philosophers¯5	20	15	possibly 2nd proposed method	0.009
beverage¯production	20	16	possibly 2nd proposed method	0.013
miling¯machine	20	16	possibly 2nd proposed method	0.013
milling	20	16	possibly 2nd proposed method	0.013
mixer mod1	20	16	possibly 2nd proposed method	0.013
dingle1	20	20	possibly 2nd proposed method	0.015
basile1	21	17	possibly 2nd proposed method	0.013
lnet p8n3	21	17	possibly 2nd proposed method	0.015
fernandez3	21	20	possibly 2nd proposed method	0.017
pn fernandez3	21	20	possibly 2nd proposed method	0.016
heiner1	22	20	possibly 2nd proposed method	0.016
chrzastowski2	22	21	possibly 2nd proposed method	0.019
vanDerAalst5	23	23	possibly 2nd proposed method	0.021
adam1	24	12	possibly 2nd proposed method	0.008
xie1	24	26	possibly 2nd proposed method	0.025
HAN _ _ _	24	40	possibly 2nd proposed method	0.035
oil separator cover alfa net	27	21	possibly 2nd proposed method	0.025

(Table A.3) cont.....

lnet_p8n2	28	17	possibly 2nd proposed method	0.018
zuberek3	29	21	possibly 2nd proposed method	0.025
alfa_net_copy_milling_quality	29	25	false 1st proposed method	0.030
oil_separator_cover_s_net	29	25	possibly 2nd proposed method	0.032
ConsistentExample	29	26	possibly 1st proposed method	0.042
zuberek4	30	21	possibly 2nd proposed method	0.027
zuberek1	30	22	possibly 2nd proposed method	0.028
well-built-alfa-net	30	25	possibly 2nd proposed method	0.035
alfa-net-copy-milling-subprocess	30	26	possibly 1st proposed method	0.035
s-net-copy-milling	31	28	possibly 2nd proposed method	0.042
crossroadSM-FPGA	32	12	possibly 2nd proposed method	0.012
alfa-net-copy-milling-synch	32	27	possibly 2nd proposed method	0.053
s-net-copy-milling	32	29	false 1st proposed method	0.052
state-space32	32	32	possibly 2nd proposed method	0.035
vanDerAalst8	34	27	possibly 2nd proposed method	0.042
lnet_p4n1	37	41	possibly 2nd proposed method	0.093
zuberek5	41	31	possibly 2nd proposed method	0.068
lnet_p7n1	41	32	possibly 2nd proposed method	0.072
state_space48	48	48	possibly 2nd proposed method	0.100
PWM_extended	49	31	possibly 2nd proposed method	0.080
lnet_p8n1	51	40	possibly 1st proposed method	0.131
cn_crr7	56	15	possibly 2nd proposed method	0.026
cn_crr10	80	21	possibly 2nd proposed method	0.064
cn_crr15	120	31	possibly 2nd proposed method	0.193
cn_crr25	200	51	possibly 2nd proposed method	0.832

Table A.4. Full results of experiments, combined proposed methods.

Name	$\|P\|$	$\|T\|$	Toolchain Bounded	Applied Methods	Runtime [ms]
prio ex	2	3	true	3rd + 2nd prop. method	0.090
semaphore	3	4	true	3rd + 2nd prop. method	0.087
coloured	4	3	true	3rd + 2nd prop. method	0.091
traffic light v2	4	3	true	3rd + 2nd prop. method	0.090
pn silva 05e	4	4	false	3rd + 1st prop. method	16.960

(Table A.4) cont.....

tank_heating	4	4	true	$3^{rd} + 2^{nd}$ prop. method	0.082
Entrance1	4	6	true	$3^{rd} + 2^{nd}$ prop. method	0.118
kovalyov92	4	6	true	$3^{rd} + 2^{nd}$ prop. method	0.089
return_books	4	6	true	$3^{rd} + 2^{nd}$ prop. method	0.095
pn_silva_05c	5	3	false	$3^{rd} + 1^{st}$ prop. method	16.998
PUMA_unloading	5	3	true	$3^{rd} + 2^{nd}$ prop. method	0.085
consumerReachability	5	4	false	$3^{rd} + 1^{st}$ prop. method	21.417
pn-silva-05b	5	4	false	$3^{rd} + 1^{st}$ prop. method	21.950
pn-silva-05f	5	4	true	$3^{rd} + 2^{nd}$ prop. method	0.104
pn-silva-04	5	6	false	$3^{rd} + 1^{st}$ prop. method	21.406
invariants-exponent-3-2	6	2	true	$3^{rd} + 2^{nd}$ prop. method	0.099
np3	6	3	true	$3^{rd} + 2^{nd}$ prop. method	0.122
gals-example	6	4	true	$3^{rd} + 2^{nd}$ prop. method	0.124
lnet_p1n1	6	4	true	$3^{rd} + 2^{nd}$ prop. method	0.148
pcncp	6	4	true	$3^{rd} + 2^{nd}$ prop. method	0.119
pn_silva_02	6	4	true	$3^{rd} + 2^{nd}$ prop. method	0.093
PUMA_loading	6	4	true	$3^{rd} + 2^{nd}$ prop. method	0.090
RHINO_loading	6	4	true	$3^{rd} + 2^{nd}$ prop. method	0.088
RHINO_unloading	6	4	true	$3^{rd} + 2^{nd}$ prop. method	0.094
lnet_p1n2	6	5	true	$3^{rd} + 2^{nd}$ prop. method	0.111
lnet_p6n3	6	5	true	$3^{rd} + 2^{nd}$ prop. method	0.106
net1	6	5	true	$3^{rd} + 2^{nd}$ prop. method	0.112
traffic lights	6	5	true	$3^{rd} + 2^{nd}$ prop. method	0.138
fig311 01	6	6	true	$3^{rd} + 2^{nd}$ prop. method	0.145
PNwD	6	6	true	$3^{rd} + 2^{nd}$ prop. method	0.152
pn desel 03	6	7	true	$3^{rd} + 2^{nd}$ prop. method	0.129
communications protocol	6	8	true	$3^{rd} + 2^{nd}$ prop. method	0.130
health care process	6	8	true	$3^{rd} + 2^{nd}$ prop. method	0.179
cncrr001	7	4	true	$3^{rd} + 2^{nd}$ prop. method	0.190
FMS main SIPN	7	4	true	$3^{rd} + 2^{nd}$ prop. method	0.180
agostini1	7	5	true	$3^{rd} + 2^{nd}$ prop. method	0.125
CNC machine	7	5	true	$3^{rd} + 2^{nd}$ prop. method	0.096
transfer	7	5	true	$3^{rd} + 1^{st}$ prop. method	24.614
voorhoeve3	7	5	true	$3^{rd} + 2^{nd}$ prop. method	0.104
net2	7	6	false	$3^{rd} + 1^{st}$ prop. method	26.420
pnbrexpl 05	7	6	true	$3^{rd} + 2^{nd}$ prop. method	0.308
two traffic lights	7	6	true	$3^{rd} + 2^{nd}$ prop. method	0.129

(Table A.4) cont.....

pn_desel_01	7	7	true	3rd + 2nd prop. method	0.100
silva1	7	7	true	3rd + 2nd prop. method	0.107
traffic_light_v1	8	3	true	3rd + 2nd prop. method	2.755
agerwala1	8	4	true	3rd + 2nd prop. method	0.170
gaubert2	8	4	true	3rd + 2nd prop. method	0.171
mixer-mod2	8	5	true	3rd + 2nd prop. method	7.390
pn-desel-02	8	5	true	3rd + 2nd prop. method	0.136
prod-cons	8	5	true	3rd + 2nd prop. method	0.143
sub-task-of-PLT-and-PMN˜	8	5	true	3rd + 2nd prop. method	0.146
bridge	8	6	true	3rd + 2nd prop. method	0.103
cp2	8	6	true	3rd + 1st prop. method	33.627
Exe5-split	8	6	true	3rd + 2nd prop. method	0.103
img-280	8	6	true	3rd + 2nd prop. method	0.120
lab5	8	6	true	3rd + 2nd prop. method	0.145
TP5-I	8	6	true	3rd + 2nd prop. method	0.107
bit-protocol	8	7	false	3rd + 1st prop. method	27.620
net4	8	7	true	3rd + 2nd prop. method	0.160
silva14	8	7	true	3rd + 2nd prop. method	0.498
elevator-2	8	8	true	3rd + 2nd prop. method	0.109
girault8	8	8	true	3rd + 2nd prop. method	0.162
net3	8	8	true	3rd + 2nd prop. method	0.163
hard case	9	3	true	3rd + 2nd prop. method	0.128
invariants exponent 3 3	9	3	true	3rd + 2nd prop. method	0.134
bridge semaphore	9	6	true	3rd + 2nd prop. method	0.215
cortadella1	9	6	true	3rd + 2nd prop. method	0.127
lnet p1n4	9	6	true	3rd + 2^{n}d prop. method	0.146
multi robot	9	6	true	3rd + 2nd prop. method	0.175
oneWayTransmissionSystem	9	6	true	3rd + 2nd prop. method	0.178
semaphore2	9	6	true	3rd + 2nd prop. method	0.142
simple production system	9	6	true	3rd + 2nd prop. method	0.211
lnet p2n3	9	7	true	3rd + 2nd prop. method	0.134
mutual exclusion	9	7	true	3rd + 2nd prop. method	0.239
miczulski1	9	8	true	3rd + 2nd prop. method	0.136
pnbrexpl 03	9	8	true	3rd + 2nd prop. method	0.280
reactor small	9	8	true	3rd + 2nd prop. method	0.136
pn silva 01	9	9	true	3rd + 2nd prop. method	0.154
medeiros1	9	13	true	3rd + 2nd prop. method	0.197

(Table A.4) cont.....

vanDerAalst6	9	13	true	$3^{rd} + 2^{nd}$ prop. method	0.521
np5	10	5	true	$3^{rd} + 2^{nd}$ prop. method	0.223
ConsistentExampleMessageView	10	6	true	$3^{rd} + 2^{nd}$ prop. method	0.225
pn_silva_05d	10	6	true	$3^{rd} + 1^{st}$ prop. method	62.794
exOR	10	7	true	$3^{rd} + 1^{st}$ prop. method	21.011
girault4	10	7	true	$3^{rd} + 2^{nd}$ prop. method	0.190
speedway	10	7	true	$3^{rd} + 2^{nd}$ prop. method	0.250
barkaoui2	10	8	true	$3^{rd} + 2^{nd}$ prop. method	0.229
credit-procedure	10	8	true	$3^{rd} + 2^{nd}$ prop. method	0.138
lnet-p5n3	10	8	true	$3^{rd} + 2^{nd}$ prop. method	0.146
medeiros2	10	8	true	$3^{rd} + 2^{nd}$ prop. method	0.255
pascal-reach	10	8	true	$3^{rd} + 2^{nd}$ prop. method	1.154
silva8	10	8	true	$3^{rd} + 1^{st}$ prop. method	22.074
jeng2	10	9	true	$3^{rd} + 2^{nd}$ prop. method	0.169
chrzastowski1	10	10	true	$3^{rd} + 2^{nd}$ prop. method	0.247
sivaraman1	10	10	true	$3^{rd} + 2^{nd}$ prop. method	0.151
franczok1	10	14	true	$3^{rd} + 2^{nd}$ prop. method	0.199
cncrr002	11	7	true	$3^{rd} + 2^{nd}$ prop. method	0.263
dataflow_computation	11	7	true	$3^{rd} + 2^{nd}$ prop. method	0.239
girault7	11	7	true	$3^{rd} + 2^{nd}$ prop. method	0.178
parallel_managing_automatic	11	7	false	$3^{rd} + 1^{st}$ prop. method	25.712
campos1	11	8	true	$3^{rd} + 2^{nd}$ prop. method	0.339
julvez 1	11	8	true	$3^{rd} + 2^{nd}$ prop. method	0.261
lnet p5n1	11	9	true	$3^{rd} + 2^{nd}$ prop. method	0.162
pnbrexpl 06	11	9	true	$3^{rd} + 2^{nd}$ prop. method	0.169
esparza1	11	10	true	$3^{rd} + 2^{nd}$ prop. method	0.329
fernandez2	11	10	true	$3^{rd} + 2^{nd}$ prop. method	0.213
rejers2	11	10	true	$3^{rd} + 2^{nd}$ prop. method	0.142
bause1	11	11	true	$3^{rd} + 1^{st}$ prop. method	30.378
oil separator cover s net v3	11	11	true	$3^{rd} + 2^{nd}$ prop. method	0.226
pnbrexpl 07	11	12	true	$3^{rd} + 2^{nd}$ prop. method	0.195
barkaoui1	11	13	true	$3^{rd} + 2^{nd}$ prop. method	0.249
li3	11	14	true	$3^{rd} + 2^{nd}$ prop. method	0.241
invariants exponent 3 4	12	4	true	$3^{rd} + 2^{nd}$ prop. method	0.141
girault1	12	6	true	$3^{rd} + 2^{nd}$ prop. method	0.354
real-life system smarthome	12	7	true	$3^{rd} + 2^{nd}$ prop. method	0.198
3carros	12	8	true	$3^{rd} + 2^{nd}$ prop. method	0.121

(Table A.4) cont.....

four_philosophers	12	8	true	$3^{rd} + 2^{nd}$ prop. method	0.323
lnet_p10n1	12	9	true	$3^{rd} + 2^{nd}$ prop. method	0.134
machine_shop	12	9	true	$3^{rd} + 1^{st}$ prop. method	7184.005
campos2	12	10	true	$3^{rd} + 2^{nd}$ prop. method	0.323
julvez_2	12	11	true	$3^{rd} + 2^{nd}$ prop. method	0.565
lnet_p6n1	12	11	true	$3^{rd} + 2^{nd}$ prop. method	0.165
bouillard1	12	12	true	$3^{rd} + 2^{nd}$ prop. method	0.252
li2	12	12	true	$3^{rd} + 2^{nd}$ prop. method	0.142
maruster3	12	13	true	$3^{rd} + 2^{nd}$ prop. method	0.170
ping1	12	14	true	$3^{rd} + 2^{nd}$ prop. method	0.266
vanDerAalst1	12	14	true	$3^{rd} + 2^{nd}$ prop. method	0.237
weijters1	12	14	true	$3^{rd} + 2^{nd}$ prop. method	0.514
Elevator01	12	17	true	$3^{rd} + 1^{st}$ prop. method	115.697
lnet_p2n2	13	7	true	$3^{rd} + 2^{nd}$ prop. method	0.293
reiseg1	13	9	true	$3^{rd} + 2^{nd}$ prop. method	0.270
ExampleStateView	13	10	true	$3^{rd} + 1^{st}$ prop. method	109.936
frame_manufact	13	10	true	$3^{rd} + 2^{nd}$ prop. method	0.210
holloway1	13	10	true	$3^{rd} + 2^{nd}$ prop. method	0.224
pnbrexpl_02	13	10	true	$3^{rd} + 2^{nd}$ prop. method	0.141
fernandez1	13	11	true	$3^{rd} + 2^{nd}$ prop. method	0.303
hee2	13	11	true	$3^{rd} + 2^{nd}$ prop. method	0.475
lnet_p5n2	13	11	true	$3^{rd} + 2^{nd}$ prop. method	0.179
pnbrexpl_08	13	11	true	$3^{rd} + 2^{nd}$ prop. method	0.180
pn_campos_silva2	13	12	true	$3^{rd} + 2^{nd}$ prop. method	4.133
pnbrexpl_15	13	12	true	$3^{rd} + 2^{nd}$ prop. method	0.253
silva4v2	13	12	true	$3^{rd} + 2^{nd}$ prop. method	0.427
voorhoeve1	13	12	true	$3^{rd} + 2^{nd}$ prop. method	0.288
board_game	13	13	false	$3^{rd} + 1^{st}$ prop. method	15.292
hack1	13	13	true	$3^{rd} + 2^{nd}$ prop. method	0.832
pn_campos_silva7	13	13	true	$3^{rd} + 2^{nd}$ prop. method	0.323
pnbrexpl_04	13	13	true	$3^{rd} + 2^{nd}$ prop. method	0.288
silva7	13	13	true	$3^{rd} + 2^{nd}$ prop. method	0.320
maruster1	13	14	true	$3^{rd} + 2^{nd}$ prop. method	0.285
esparza3	13	15	true	$3^{rd} + 2^{nd}$ prop. method	0.393
girault3	14	9	true	$3^{rd} + 2^{nd}$ prop. method	0.951
philosophers_2_rev_2	14	10	true	$3^{rd} + 2^{nd}$ prop. method	0.469
philosophers_2	14	10	true	$3^{rd} + 2^{nd}$ prop. method	0.479

(Table A.4) cont.....

pn_silva_03	14	10	true	$3^{rd} + 2^{nd}$ prop. method	0.326
rejers1	14	11	true	$3^{rd} + 2^{nd}$ prop. method	0.140
hee1	14	12	true	$3^{rd} + 2^{nd}$ prop. method	0.573
voorhoeve2	14	12	true	$3^{rd} + 2^{nd}$ prop. method	0.327
pnbrexpl-12	14	13	true	$3^{rd} + 2^{nd}$ prop. method	0.255
vanDerAalst3	14	13	true	$3^{rd} + 2^{nd}$ prop. method	0.146
vanDerAalst7	14	13	true	$3^{rd} + 2^{nd}$ prop. method	0.573
jeng1	14	16	true	$3^{rd} + 2^{nd}$ prop. method	0.318
lnet-p6n2	14	16	false	$3^{rd} + 1^{st}$ prop. method	49.145
lnet-p9n1	14	16	true	$3^{rd} + 2^{nd}$ prop. method	0.272
desel1	15	9	true	$3^{rd} + 2^{nd}$ prop. method	0.169
dining-philosophers	15	10	false	$3^{rd} + 1^{st}$ prop. method	45.730
campos3	15	11	true	$3^{rd} + 2^{nd}$ prop. method	0.572
oil-separator-cover-s-net-v2	15	11	true	$3^{rd} + 2^{nd}$ prop. method	3.057
IEC	15	12	true	$3^{rd} + 2^{nd}$ prop. method	0.220
esparza2	15	13	true	$3^{rd} + 2^{nd}$ prop. method	0.552
lasire1	15	13	true	$3^{rd} + 2^{nd}$ prop. method	0.187
lnet-p8n4	15	16	true	$3^{rd} + 2^{nd}$ prop. method	0.407
dongen1	15	17	true	$3^{rd} + 2^{nd}$ prop. method	0.329
2pusher	15	18	false	$3^{rd} + 1^{st}$ prop. method	68.606
gaubert1	16	8	true	$3^{rd} + 2^{nd}$ prop. method	0.495
silva5	16	8	true	$3^{rd} + 2^{nd}$ prop. method	0.551
girault2	16	10	true	$3^{rd} + 2^{nd}$ prop. method	0.544
brenner1	16	12	true	$3^{rd} + 2^{nd}$ prop. method	0.529
beverage production part2	16	13	true	$3^{rd} + 2^{nd}$ prop. method	0.281
beverage	16	13	true	$3^{rd} + 2^{nd}$ prop. method	0.234
maruster4	16	13	true	$3^{rd} + 2^{nd}$ prop. method	0.223
mixer one cup	16	13	true	$3^{rd} + 2^{nd}$ prop. method	10.055
pnbrexpl 09	16	13	true	$3^{rd} + 2^{nd}$ prop. method	0.267
chiola1	16	15	true	$3^{rd} + 2^{nd}$ prop. method	0.582
state space16	16	16	true	$3^{rd} + 2^{nd}$ prop. method	2.175
handling of incoming order	17	14	true	$3^{rd} + 2^{nd}$ prop. method	0.412
eshuis1	17	15	true	$3^{rd} + 2^{nd}$ prop. method	0.337
automation3	17	16	true	$3^{rd} + 1^{st}$ prop. method	66.741
s net frame manufact	17	16	true	$3^{rd} + 2^{nd}$ prop. method	3.839
li1	17	17	true	$3^{rd} + 2^{nd}$ prop. method	0.262
pn campos silva6	18	11	true	$3^{rd} + 2^{nd}$ prop. method	0.597

(Table A.4) cont.....

barkaoui3	18	13	true	$3^{rd} + 2^{nd}$ prop. method	0.786
balduzzi1	18	16	true	$3^{rd} + 1^{st}$ prop. method	31.703
vanDerAalst4	18	16	true	$3^{rd} + 2^{nd}$ prop. method	0.354
kavi1	18	20	true	$3^{rd} + 2^{nd}$ prop. method	0.516
PWM-patterns	19	11	true	$3^{rd} + 2^{nd}$ prop. method	0.320
hulgaard1	19	12	true	$3^{rd} + 2^{nd}$ prop. method	0.981
mixer	19	15	true	$3^{rd} + 2^{nd}$ prop. method	0.331
s-net-frame-manufact	19	18	true	$3^{rd} + 2^{nd}$ prop. method	0.346
maruster2	20	14	true	$3^{rd} + 2^{nd}$ prop. method	0.573
holliday1	20	15	true	$3^{rd} + 2^{nd}$ prop. method	0.641
philosophers-5	20	15	true	$3^{rd} + 2^{nd}$ prop. method	0.564
beverage-production	20	16	true	$3^{rd} + 2^{nd}$ prop. method	0.359
miling-machine	20	16	true	$3^{rd} + 2^{nd}$ prop. method	0.206
milling	20	16	true	$3^{rd} + 2^{nd}$ prop. method	0.200
mixer-mod1	20	16	true	$3^{rd} + 2^{nd}$ prop. method	0.384
dingle1	20	20	true	$3^{rd} + 2^{nd}$ prop. method	0.910
basile1	21	17	true	$3^{rd} + 2^{nd}$ prop. method	0.435
lnet-p8n3	21	17	true	$3^{rd} + 2^{nd}$ prop. method	0.343
fernandez3	21	20	true	$3^{rd} + 2^{nd}$ prop. method	0.691
pn-fernandez3	21	20	true	$3^{rd} + 2^{nd}$ prop. method	0.943
heiner1	22	20	true	$3^{rd} + 2^{nd}$ prop. method	0.916
chrzastowski2	22	21	true	$3^{rd} + 2^{nd}$ prop. method	0.242
vanDerAalst5	23	23	true	$3^{rd} + 2^{nd}$ prop. method	0.747
adam1	24	12	true	$3^{rd} + 2^{nd}$ prop. method	1.892
xie1	24	26	true	$3^{rd} + 2^{nd}$ prop. method	0.656
HAN	24	40	true	$3^{rd} + 2^{nd}$ prop. method	1.270
oil separator cover alfa net	27	21	true	$3^{rd} + 2^{nd}$ prop. method	0.493
lnet p8n2	28	17	true	$3^{rd} + 2^{nd}$ prop. method	4.609
zuberek3	29	21	true	$3^{rd} + 2^{nd}$ prop. method	0.705
alfa net copy milling quality	29	25	false	$3^{rd} + 1^{st}$ prop. method	55.725
oil separator cover s net	29	25	true	$3^{rd} + 2^{nd}$ prop. method	181.018
ConsistentExample	29	26	true	$3^{rd} + 1^{st}$ prop. method	0.601
zuberek4	30	21	true	$3^{rd} + 2^{nd}$ prop. method	1.351
zuberek1	30	22	true	$3^{rd} + 2^{nd}$ prop. method	1.675
well built alfa net	30	25	true	$3^{rd} + 2^{nd}$ prop. method	0.572
alfa net copy milling subprocess	30	26	true	$3^{rd} + 1^{st}$ prop. method	48.734
s net copy milling	31	28	true	$3^{rd} + 2^{nd}$ prop. method	0.916

(Table A.4) cont.....

crossroadSM_FPGA	32	12	true	$3^{rd} + 2^{nd}$ prop. method	1.461
alfa_net_copy_milling_synch	32	27	true	$3^{rd} + 2^{nd}$ prop. method	0.697
s_net_copy_milling	32	29	false	$3^{rd} + 1^{st}$ prop. method	142.583
state_space32	32	32	true	$3^{rd} + 2^{nd}$ prop. method	2.289
vanDerAalst8	34	27	true	$3^{rd} + 2^{nd}$ prop. method	4.827
lnet-p4n1	37	41	true	$3^{rd} + 2^{nd}$ prop. method	4.782
zuberek5	41	31	true	$3^{rd} + 2^{nd}$ prop. method	2.459
lnet-p7n1	41	32	true	$3^{rd} + 2^{nd}$ prop. method	1.511
state-space48	48	48	true	$3^{rd} + 2^{nd}$ prop. method	8.444
PWM-extended	49	31	true	$3^{rd} + 2^{nd}$ prop. method	17.922
lnet-p8n1	51	40	true	$3^{rd} + 1^{st}$ prop. method	42494.837
cn_crr7	56	15	true	$3^{rd} + 2^{nd}$ prop. method	3.858
cn_crr10	80	21	true	$3^{rd} + 2^{nd}$ prop. method	12.846
cn_crr15	120	31	true	$3^{rd} + 2^{nd}$ prop. method	37.924
cn_crr25	200	51	true	$3^{rd} + 2^{nd}$ prop. method	207.225

A.2 SAFENESS

Table A.5. Full results of experiments, state space-based algorithms.

Name	|P|	|T|	Reference Method Safe Runtime[ms]	4th Proposed Method Safe Runtime[ms]
prio ex	2	3 true	19.0780 true	15.5175
semaphore	3	4 true	19.0501 true	15.6182
coloured	4	3 true	25.1824 true	16.2993
traffic_light_v2	4	3 true	22.1925 true	15.9034
pn_silva_05e	4	4 false	22.9779 false	13.9798
tank_heating	4	4 true	19.9551 true	13.0749
Entrance1	4	6 true	23.2347 true	16.3483
kovalyov92	4	6 true	20.1264 true	15.3583
return_books	4	6 true	28.4905 true	14.1438
pn_silva_05c	5	3 false	21.3463 false	13.7149
PUMA-unloading	5	3 true	20.9865 true	18.0956

(Table A.5) cont.....

consumerReachability	5	4 false	26.4103 false	14.5373
pn_silva_05b	5	4 false	32.2826 false	15.6270
pn_silva_05f	5	4 true	23.3149 true	14.4269
pn_silva_04	5	6 false	41.1586 false	18.0003
invariants_exponent_3_2	6	2 true	24.2562 true	17.9393
np3	6	3 true	25.4518 true	16.1491
gals-example	6	4 true	28.6123 true	18.6427
lnet-p1n1	6	4 true	24.3566 true	15.1327
pcncp	6	4 true	21.9622 true	16.2125
pn-silva-02	6	4 true	19.6898 true	15.0080
PUMA-loading	6	4 true	20.7406 true	14.5736
RHINO-loading	6	4 true	22.6433 true	15.5960
RHINO-unloading	6	4 true	23.7006 true	14.7157
lnet-p1n2	6	5 true	22.7569 true	14.4633
lnet-p6n3	6	5 true	25.7643 true	15.7221
net1	6	5 true	29.2881 true	16.3986
traffic_lights	6	5 true	71.1782 true	14.8824
fig311_01	6	6 true	23.7371 true	20.7076
PNwD	6	6 false	44.1532 false	5.0541
pn_desel_03	6	7 true	23.9147 true	15.0002
communications_protocol	6	8 true	23.6780 true	15.5404
health_care_process	6	8 true	26.3826 true	31.2681
cncrr001	7	4 true	18.5514 true	16.3655
FMS_main_SIPN	7	4 true	24.6493 true	15.9028
agostini1	7	5 true	17.7040 true	14.1424
CNC machine	7	5 true	17.4798 true	17.6813
transfer	7	5 false	71.8959 false	19.4545
voorhoeve3	7	5 true	23.5822 true	19.2303
net2 _	7	6 false	43.1173 false	22.8879
pnbrexpl 05	7	6 true	22.7454 true	16.0845
two traffic lights	7	6 true	25.3653 true	15.7590
pn desel 01	7	7 true	24.8940 true	18.4418
silva1_ _	7	7 true	23.6710 true	17.2683
traffic light v1	8	3 true	21.4085 true	16.6392
agerwala1	8	4 true	16.4393 true	15.5919
gaubert2	8	4 true	25.4083 true	15.6056
mixer mod2	8	5 true	24.4189 true	16.0223

(Table A.5) cont.....

pn_desel_02	8	5 true	30.4172 true	18.0021
prod_cons	8	5 true	35.2817 true	25.1723
sub-task_of_PLT_and_PMN~	8	5 true	54.7568 true	15.1413
bridge	8	6 true	24.2906 true	18.2040
cp2	8	6 true	34.2382 true	19.6579
Exe5-split	8	6 true	23.0403 true	18.7990
img-280	8	6 true	36.9830 true	21.2906
lab5	8	6 true	25.9867 true	17.9450
TP5-I	8	6 true	64.7127 true	17.1282
bit-protocol	8	7 false	48.2556 false	27.0238
net4	8	7 true	26.2908 true	24.8691
silva14	8	7 true	27.5944 true	18.0024
elevator-2	8	8 true	27.9761 true	18.3702
girault8	8	8 true	43.4841 true	30.9473
net3	8	8 true	30.9293 true	17.4025
hard_case	9	3 true	24.4431 true	35.6013
invariants_exponent_3_3	9	3 true	24.4752 true	16.8175
bridge_semaphore	9	6 true	24.4369 true	17.8110
cortadella1	9	6 true	24.9902 true	18.1420
lnet_p1n4	9	6 true	24.1594 true	17.2365
multi_robot	9	6 true	29.1748 true	21.1245
oneWayTransmissionSystem	9	6 true	35.3004 true	21.5588
semaphore2	9	6 true	23.8924 true	16.7402
simple production system	9	6 true	33.8013 true	20.7964
lnet p2n3	9	7 true	23.5703 true	17.1914
mutual exclusion	9	7 true	33.7939 true	18.2336
miczulski1	9	8 true	27.4469 true	19.2999
pnbrexpl 03	9	8 true	28.9208 true	18.2586
reactor small	9	8 true	28.2865 true	17.7473
pn silva 01	9	9 true	23.0759 true	15.3197
medeiros1	9	13 true	27.5173 true	16.6554
vanDerAalst6	9	13 true	61.7860 true	16.8938
np5	10	5 true	21.8705 true	16.6962
ConsistentExampleMessageView	10	6 true	27.3633 true	19.9430
pn silva 05d	10	6 true	23.3053 true	16.8381
exOR	10	7 true	19.8099 true	1157425.0000
girault4	10	7 true	42.2787 true	29.8964

(Table A.5) cont.....

speedway	10	7 true	24.4475 true	16.2454
barkaoui2	10	8 true	19.3042 true	16.5852
credit_procedure	10	8 true	37.2567 true	22.0580
lnet_p5n3	10	8 true	38.5621 true	23.7509
medeiros2	10	8 true	29.7437 true	19.3634
pascal_reach	10	8 true	33.7933 true	19.1851
silva8	10	8 true	23.9648 true	12.8131
jeng2	10	9 true	28.1174 true	21.5514
chrzastowski1	10	10 true	28.6589 true	29.6647
sivaraman1	10	10 true	30.2501 true	20.3373
franczok1	10	14 true	32.7726 true	18.3673
cncrr002	11	7 true	44.9520 true	43.9512
dataflow¯computation	11	7 true	37.5029 true	23.9803
girault7	11	7 true	28.7862 true	22.5678
parallel¯managing¯automatic˜	11	7 false	43.6825 false	23.1102
campos1	11	8 true	34.0170 true	49.6388
julvez¯1	11	8 true	76.1174 true	52.6802
lnet¯p5n1	11	9 true	45.7711 true	26.0459
pnbrexpl¯06	11	9 true	29.7195 true	18.2188
esparza1	11	10 true	53.2915 true	37.2374
fernandez2	11	10 true	25.2160 true	19.1360
rejers2	11	10 true	26.6619 true	19.8361
bause1	11	11 true	28.1506 true	23.5252
oil separator cover s net v3	11	11 true	31.3567 true	21.7072
pnbrexpl 07	11	12 true	62.2016 true	42.2002
barkaoui1	11	13 true	31.7694 true	27.0118
li3	11	14 true	38.1803 true	24.0387
invariants exponent 3 4	12	4 true	21.8958 true	16.5507
girault1	12	6 true	40.1864 true	18.5023
real-life system smarthome˜	12	7 true	37.3951 true	24.4497
3carros	12	8 true	27.9686 true	29.1537
four_philosophers	12	8 true	32.9398 true	23.0046
lnet p10n1	12	9 true	78.4702 true	48.0305
machine shop	12	9 true	1153650.0000 true	5.1309
campos2	12	10 true	24.6993 true	24.3451
julvez 2	12	11 true	34.7127 true	18.5087
lnet p6n1	12	11 true	48.5221 true	29.8166

(Table A.5) cont.....

bouillard1	12	12 true	49.1283 true	40.0612
li2	12	12 true	43.0491 true	27.1457
maruster3	12	13 true	28.5800 true	19.7536
ping1	12	14 true	35.1854 true	27.3093
vanDerAalst1	12	14 true	79.9651 true	26.9047
weijters1	12	14 true	80.0302 true	59.1239
Elevator01	12	17 true	75.6675 true	64.3786
lnet-p2n2	13	7 true	26.7647 true	21.5208
reiseg1	13	9 true	42.2186 true	26.8663
ExampleStateView	13	10 false	78.4403 false	44.1234
frame-manufact	13	10 true	43.2207 true	28.1516
holloway1	13	10 true	59.2947 true	43.6527
pnbrexpl-02	13	10 true	45.9803 true	30.7972
fernandez1	13	11 true	25.3959 true	18.8408
hee2	13	11 true	30.6845 true	37.5002
lnet¯p5n2	13	11 true	54.1923 true	35.0413
pnbrexpl¯08	13	11 true	38.3248 true	24.6866
pn¯campos¯silva2	13	12 true	72.3560 true	51.4288
pnbrexpl¯15	13	12 true	50.0065 true	40.8692
silva4v2	13	12 true	69.5506 true	52.6372
voorhoeve1	13	12 true	112.3877 true	44.4445
board¯game	13	13 false	21.7698 false	14.3919
hack1	13	13 true	66.4891 true	53.4076
pn campos silva7	13	13 true	45.6876 true	32.2327
pnbrexpl 04	13	13 true	43.5901 true	31.0588
silva7	13	13 true	44.8462 true	31.5231
maruster1	13	14 true	43.7007 true	29.5956
esparza3	13	15 true	68.2389 true	51.5460
girault3	14	9 true	38.2862 true	25.9363
philosophers_2_	14	10 true	45.2491 true	34.3534
philosophers 2 rev 2	14	10 true	48.3218 true	35.4309
pn silva 03	14	10 true	60.4660 true	42.1453
rejers1	14	11 true	74.0095 true	47.3889
hee1	14	12 true	40.9226 true	56.9268
voorhoeve2	14	12 true	101.0768 true	57.8930
pnbrexpl 12	14	13 true	50.3703 true	43.1007
vanDerAalst3	14	13 true	82.1756 true	29.7832

(Table A.5) cont.....

vanDerAalst7	14	13 true	122.2586 true	46.1532
jeng1	14	16 true	47.0792 true	34.6446
lnet_p6n2	14	16 false	712.1394 false	63.5090
lnet_p9n1	14	16 true	63.3058 true	46.1819
desel1	15	9 true	185.3470 true	127.7411
dining_philosophers	15	10 false	81.4642 false	48.5635
campos3	15	11 true	62.2191 true	53.1786
oil-separator-cover-s-net-v2	15	11 true	31.7127 true	19.6126
IEC	15	12 true	50.9404 true	32.4745
esparza2	15	13 true	58.5960 true	41.0707
lasire1	15	13 true	41.4052 true	21.0735
lnet-p8n4	15	16 true	75.1005 true	63.8796
dongen1	15	17 true	40.4876 true	32.7116
2pusher	15	18 false	111.1739 false	60.3173
gaubert1	16	8 true	34.4179 true	20.2161
silva5	16	8 true	52.8754 true	36.1776
girault2	16	10 true	49.6675 true	31.3195
brenner1	16	12 false	50702.5932 false	1047.3681
beverage	16	13 true	50.7734 true	44.4248
beverage_production_part2	16	13 true	48.5905 true	42.0991
maruster4	16	13 true	71.1404 true	54.7766
mixer_one_cup	16	13 true	66.9388 true	42.0077
pnbrexpl_09	16	13 true	64.9904 true	42.7809
chiola1	16	15 true	34.2005 true	27.5894
state space16	16	16 true	1493.9547 true	1435.8302
handling of incoming order	17	14 true	42.2454 true	29.6139
eshuis1	17	15 true	50.8159 true	30.3682
automation3	17	16 true	60.3088 true	72.3349
s net frame manufact quality	17	16 true	91.6302 true	29.2645
li1	17	17 true	57.5485 true	44.2736
pn campos silva6	18	11 true	9.0868 true	4.9494
barkaoui3	18	13 true	63.8518 true	49.5238
balduzzi1	18	16 true	27.0226 true	23.2883
vanDerAalst4	18	16 true	240.5476 true	77.2516
kavi1	18	20 true	169.3538 true	123.0547
PWM patterns	19	11 true	45.1191 true	28.2835
hulgaard1	19	12 true	41.0423 true	27.2892

(Table A.5) cont.....

mixer	19	15 true	63.0277 true	45.5941
s_net_frame_manufact_quality˜	19	18 true	95.0531 true	41.6006
maruster2	20	14 true	65.5881 true	51.6015
holliday1	20	15 true	295.4204 true	183.8872
philosophers_5	20	15 true	105.0783 true	75.8251
beverage_production	20	16 true	57.7263 true	46.4452
miling_machine	20	16 true	89.0111 true	58.0210
milling	20	16 true	78.1119 true	60.3329
mixer_mod1	20	16 true	65.7985 true	46.1381
dingle1	20	20 true	45.4625 true	30.8079
basile1	21	17 true	106.6939 true	86.7165
lnet_p8n3	21	17 true	94.4362 true	67.8008
fernandez3	21	20 true	43.3925 true	30.9028
pn_fernandez3	21	20 true	48.2607 true	27.8295
heiner1	22	20 true	130.7727 true	128.0666
chrzastowski2	22	21 true	43.6154 true	41.7286
vanDerAalst5	23	23 true	115.9249 true	57.5321
adam1	24	12 true	54.2918 true	56.6285
xie1	24	26 true	292.7697 true	103.4013
HAN	24	40 true	50231.3384 true	53669.2779
oil_separator_cover_alfa_net	27	21 true	57.0549 true	41.8627
lnet_p8n2	28	17 true	99.5833 true	71.4567
zuberek3	29	21 true	692.0424 true	188.9980
alfa_net_copy_milling˜_quality	29	25 false	78.3404 false	50.3740
ConsistentExample	29	25 false	91.6349 false	45.2747
oil_separator_cover_s_net	29	25 true	66.2475 true	82.6780
zuberek4	30	21 true	631.3786 true	192.1192
zuberek1	30	22 true	281.9953 true	102.7223
well_built_alfa_net˜_subprocess	30	25 true	531.8516 true	141.2623
alfa_net_copy˜_subprocess	30	26 true	49.3245 true	43.1146
s_net_copy_milling˜_subprocess	31	28 true	433.4854 true	183.4345
crossroadSM_FPGA	32	12 true	43.8308 true	46.3888
alfa_net_copy_milling˜_synchr	32	27 true	238.9907 true	173.8360
s_net_copy_milling˜_quality	32	29 false	41407.1404 false	145.0464
state_space32	32	32 n/a	n/a n/a	n/a
vanDerAalst8	34	27 true	352.7851 true	119.9650
lnet_p4n1	37	41 true	56.5367 true	48.7006

(Table A.5) cont.....

zuberek5	41	31	true	557.5064	true	146.1737
lnet_p7n1	41	32	true	127849.4493	true	132817.9162
state_space48	48	48	n/a	n/a	n/a	n/a
PWM_extended	49	31	true	398.9832	true	274.9435
lnet_p8n1	51	40	true	20157.0578	true	22438.7556
cn_crr7	56	15	true	93697.7373	true	96224.1716
cn_crr10	80	21	n/a	n/a	n/a	n/a
cn_crr15	120	31	n/a	n/a	n/a	n/a
cn_crr25	200	51	n/a	n/a	n/a	n/a

Table A.6. Full results of experiments, linear algebra-based algorithms.

Name	$\|P\|$	$\|T\|$	**Reference Method Reference Method**		**5^{ss} Proposed Method Covered Runtime[ms]**	
prio_ex	2	3	true	0.025	true	0.639
semaphore	3	4	true	0.059	true	0.168
coloured	4	3	true	0.050	true	0.094
traffic_light_v2	4	3	true	0.033	true	0.088
pn_silva_05e	4	4	false	0.067	false	0.090
tank_heating	4	4	true	0.069	true	0.099
Entrance1	4	6	true	0.024	true	0.127
kovalyov92	4	6	true	0.026	true	0.096
return-books	4	6	true	0.023	true	0.116
pn-silva-05c	5	3	false	0.045	false	0.101
PUMA-unloading	5	3	true	0.088	true	0.083
consumerReachability	5	4	false	0.081	false	0.113
pn_silva_05b	5	4	false	0.066	false	0.121
pn_silva_05f	5	4	true	0.028	true	0.115
pn_silva_04	5	6	false	0.026	false	0.120
invariants_exponent_3_2	6	2	true	0.044	true	0.090
np3	6	3	true	0.080	true	0.115
gals-example	6	4	true	0.030	true	0.141
lnet_p1n1	6	4	true	0.036	true	0.120
pcncp	6	4	true	0.076	true	0.167
pn_silva_02	6	4	true	0.046	true	0.093

(Table A.6) cont.....

PUMA_loading	6	4	true	0.059	true	0.242
RHINO_loading	6	4	true	0.142	true	0.086
RHINO_unloading	6	4	true	0.073	true	0.199
lnet_p1n2	6	5	true	0.152	true	0.118
lnet_p6n3	6	5	true	0.052	true	0.107
net1	6	5	true	0.120	true	0.123
traffic-lights	6	5	true	0.090	true	0.151
fig311-01	6	6	true	0.032	true	0.134
PNwD	6	6	false	0.036	false	0.112
pn-desel-03	6	7	true	0.035	true	0.128
communications-protocol	6	8	true	0.029	true	0.132
health-care-process	6	8	true	0.063	true	0.178
cncrr001	7	4	true	0.043	true	0.165
FMS-main-SIPN	7	4	true	0.076	true	0.131
agostini1	7	5	true	0.047	true	0.109
CNC-machine	7	5	true	0.035	true	0.086
transfer	7	5	false	0.039	false	2.895
voorhoeve3	7	5	true	0.186	true	0.097
net2	7	6	false	0.056	false	0.110
pnbrexpl-05	7	6	true	0.068	true	0.253
two-traffic-lights	7	6	true	0.066	true	0.136
pn-desel-01	7	7	true	0.041	true	0.107
silva1	7	7	true	0.088	true	0.107
traffic light v1	8	3	true	0.068	true	0.187
agerwala1	8	4	true	0.044	true	0.229
gaubert2	8	4	true	0.062	true	0.206
mixer mod2	8	5	true	0.040	true	0.226
pn desel 02	8	5	true	0.044	true	0.135
prod cons	8	5	true	0.042	true	0.148
sub-task of PLT and PMN	8	5	true	0.041	true	0.152
bridge	8	6	true	0.062	true	0.104
cp2	8	6	false	0.080	false	0.162
Exe5 split	8	6	true	0.069	true	0.227
img 280	8	6	true	0.044	true	0.127
lab5	8	6	true	0.045	true	0.133
TP5 I	8	6	true	0.049	true	0.100
bit protocol	8	7	false	0.042	false	0.178

(Table A.6) cont.....

net4	8	7	true	0.125	true	0.279
silva14	8	7	true	0.054	true	0.136
elevator_2	8	8	true	0.034	true	0.106
girault8	8	8	true	0.047	true	0.181
net3	8	8	true	0.042	true	0.133
hard-case	9	3	true	0.225	true	0.138
invariants-exponent-3-3	9	3	true	0.111	true	0.140
bridge-semaphore	9	6	true	0.072	true	0.229
cortadella1	9	6	true	0.068	true	0.170
lnet-p1n4	9	6	true	0.075	true	0.152
multi-robot	9	6	true	0.055	true	0.181
oneWayTransmissionSystem	9	6	true	0.153	true	0.178
semaphore2	9	6	true	0.089	true	0.161
simple_production_system	9	6	true	0.096	true	0.179
lnet_p2n3	9	7	true	0.211	true	0.134
mutual_exclusion	9	7	true	0.047	true	0.220
miczulski1	9	8	true	0.058	true	0.138
pnbrexpl_03	9	8	true	0.115	true	0.148
reactor_small	9	8	true	0.139	true	0.147
pn_silva_01	9	9	true	0.066	true	0.166
medeiros1	9	13	true	0.051	true	0.218
vanDerAalst6	9	13	true	0.060	true	0.214
np5	10	5	true	0.089	true	0.675
ConsistentExampleMessageView	10	6	true	0.137	true	0.271
pn silva 05d	10	6	false	0.094	false	0.263
exOR	10	7	false	0.068	false	0.289
girault4	10	7	true	0.056	true	0.220
speedway	10	7	true	0.509	true	0.204
barkaoui2	10	8	true	0.067	true	0.233
credit procedure	10	8	true	0.049	true	0.213
lnet p5n3	10	8	true	0.231	true	0.147
medeiros2	10	8	true	0.069	true	0.192
pascal reach	10	8	true	0.084	true	0.251
silva8	10	8	false	0.070	false	0.197
jeng2	10	9	true	0.084	true	0.177
chrzastowski1	10	10	true	0.048	true	0.216
sivaraman1	10	10	true	0.049	true	0.157

(Table A.6) cont.....

franczok1	10	14	true	0.157	true	0.203
cncrr002	11	7	true	0.078	true	0.273
dataflow_computation	11	7	true	0.069	true	0.244
girault7	11	7	true	0.116	true	0.165
parallel_managing_automatic_devices	11	7	false	0.080	false	0.232
campos1	11	8	true	2.461	true	0.269
julvez-1	11	8	true	0.068	true	0.231
lnet-p5n1	11	9	true	0.055	true	0.188
pnbrexpl-06	11	9	true	0.093	true	0.133
esparza1	11	10	true	0.057	true	0.359
fernandez2	11	10	true	0.059	true	0.238
rejers2	11	10	true	0.087	true	0.144
bause1	11	11	false	0.058	false	0.277
oil¯separator¯cover¯s¯net¯v3	11	11	true	0.075	true	0.232
pnbrexpl¯07	11	12	true	0.103	true	0.192
barkaoui1	11	13	true	0.065	true	0.237
li3	11	14	true	0.093	true	0.231
invariants¯exponent¯3¯4	12	4	true	0.671	true	0.183
girault1	12	6	true	0.094	true	0.347
real-life¯system¯smarthome˜	12	7	true	0.094	true	0.186
3carros	12	8	true	0.094	true	0.179
four_philosophers	12	8	true	0.097	true	0.277
lnet p10n1	12	9	true	0.075	true	0.225
machine shop	12	9	false	0.175	false	0.333
campos2	12	10	true	0.068	true	0.360
julvez 2	12	11	true	0.085	true	0.224
lnet p6n1	12	11	true	0.077	true	0.166
bouillard1	12	12	true	0.081	true	0.284
li2	12	12	true	0.069	true	0.142
maruster3	12	13	true	0.156	true	0.184
ping1	12	14	true	0.088	true	0.279
vanDerAalst1	12	14	true	0.095	true	0.292
weijters1	12	14	true	0.173	true	0.279
Elevator01	12	17	false	0.090	false	0.386
lnet p2n2	13	7	true	0.142	true	0.289
reiseg1	13	9	true	0.103	true	0.303
ExampleStateView	13	10	false	0.073	false	0.340

(Table A.6) cont.....

frame_manufact	13	10	true	0.191	true	0.141
holloway1	13	10	true	0.942	true	0.238
pnbrexpl_02	13	10	true	0.085	true	0.138
fernandez1	13	11	true	0.161	true	0.395
hee2	13	11	true	0.170	true	0.482
lnet-p5n2	13	11	true	0.230	true	0.220
pnbrexpl-08	13	11	true	0.099	true	0.177
pn-campos-silva2	13	12	true	0.082	true	0.460
pnbrexpl-15	13	12	true	0.278	true	0.230
silva4v2	13	12	true	0.076	true	0.436
voorhoeve1	13	12	true	0.216	true	0.339
board-game	13	13	false	0.107	false	0.345
hack1	13	13	true	0.103	true	0.367
pn-campos-silva7	13	13	true	0.379	true	0.349
pnbrexpl-04	13	13	true	0.134	true	0.369
silva7	13	13	true	0.119	true	0.370
maruster1	13	14	true	0.133	true	0.273
esparza3	13	15	true	0.144	true	0.413
girault3	14	9	true	0.322	true	0.447
philosophers_2	14	10	true	0.110	true	0.491
philosophers_2_rev_2	14	10	true	0.115	true	0.541
pn_silva_03	14	10	true	0.295	true	0.351
rejers1	14	11	true	0.367	true	0.136
hee1	14	12	true	0.489	true	0.609
voorhoeve2	14	12	true	0.439	true	0.357
pnbrexpl 12	14	13	true	0.093	true	0.306
vanDerAalst3	14	13	true	0.130	true	0.146
vanDerAalst7	14	13	true	0.070	true	0.207
jeng1	14	16	true	0.063	true	0.310
lnet_p6n2	14	16	false	0.087	false	0.243
lnet p9n1	14	16	true	0.206	true	0.301
desel1	15	9	true	0.064	true	0.156
dining philosophers	15	10	false	0.071	false	0.520
campos3	15	11	true	0.134	true	0.594
oil separator cover s net v2	15	11	true	0.394	true	0.328
IEC	15	12	true	0.114	true	0.184
esparza2	15	13	true	2.282	true	0.658

(Table A.6) cont.....

lasire1	15	13	true	0.123	true	0.209
lnet_p8n4	15	16	true	0.084	true	0.527
dongen1	15	17	true	0.070	true	0.422
2pusher	15	18	false	0.114	false	0.542
gaubert1	16	8	true	0.270	true	0.487
silva5	16	8	true	1.449	true	0.667
girault2	16	10	true	0.174	true	0.606
brenner1	16	12	false	0.098	false	0.495
beverage	16	13	true	0.097	true	0.231
beverage-production-part2	16	13	true	0.088	true	0.232
maruster4	16	13	true	0.169	true	0.229
mixer¯one¯cup	16	13	true	0.080	true	0.218
pnbrexpl¯09	16	13	true	0.234	true	0.275
chiola1	16	15	true	0.079	true	0.636
state¯space16	16	16	true	0.102	true	0.829
handling¯of¯incoming¯order	17	14	true	0.090	true	0.376
eshuis1	17	15	true	0.133	true	0.395
automation3	17	16	false	0.086	false	0.833
s¯net¯frame¯manufact¯quality˜	17	16	true	0.130	true	0.225
li1	17	17	true	0.548	true	0.247
pn¯campos¯silva6	18	11	false	0.146	false	0.699
barkaoui3	18	13	true	0.105	true	0.827
balduzzi1	18	16	false	0.110	false	0.766
vanDerAalst4	18	16	true	0.153	true	0.350
kavi1 _	18	20	true	0.227	true	0.516
PWM patterns	19	11	true	0.296	true	0.344
hulgaard1	19	12	true	11.882	true	1.194
mixer _ _	19	15	true	0.109	true	0.324
s net frame manufact quality˜	19	18	true	0.132	true	0.344
maruster2	20	14	true	0.166	true	0.569
holliday1 _	20	15	true	0.283	true	0.754
philosophers 5	20	15	true	0.128	true	0.625
beverage production	20	16	true	0.161	true	0.364
miling machine	20	16	true	0.142	true	0.200
milling	20	16	true	0.170	true	0.313
mixer mod1	20	16	true	0.141	true	1.058
dingle1	20	20	true	0.103	true	0.965

(Table A.6) cont.....

basile1	21	17	true	0.151	true	0.431
lnet_p8n3	21	17	true	0.174	true	0.298
fernandez3	21	20	true	0.229	true	1.379
pn_fernandez3	21	20	true	0.206	true	0.711
heiner1	22	20	true	0.143	true	1.149
chrzastowski2	22	21	true	0.107	true	0.214
vanDerAalst5	23	23	true	0.178	true	0.656
adam1	24	12	true	1.829	true	3.453
xie1	24	26	true	1.289	true	0.636
HAN	24	40	true	0.105	true	1.719
oil¯separator¯cover¯alfa¯net	27	21	true	1.249	true	0.790
lnet¯p8n2	28	17	true	0.429	true	1.670
zuberek3	29	21	true	0.575	true	0.708
alfa¯net¯copy¯milling˜¯quality	29	25	false	0.384	false	0.642
ConsistentExample	29	25	false	56 220.855	false	934.609
oil¯separator¯cover¯s¯net	29	25	true	0.413	true	0.545
zuberek4	30	21	true	1.944	true	0.945
zuberek1	30	22	true	12.602	true	3.973
well¯built¯alfa¯net˜¯subprocess	30	25	true	0.267	true	9.457
alfa¯net copy¯milling¯subprocess	30	26	false	0.368	false	0.586
s¯net copy¯milling˜ subprocess	31	28	true	0.521	true	0.635
crossroadSM¯FPGA	32	12	true	891 401.325	true	1.577
alfa net copy milling˜ synchr	32	27	true	0.367	true	1.527
s net_copy milling˜ quality	32	29	false	0.444	false	0.642
state space32	32	32	true	0.282	true	16.723
vanDerAalst8	34	27	true	0.423	true	8.872
lnet p4n1	37	41	true	0.172	true	3.571
zuberek5	41	31	n/a	n/a	true	2.726
lnet p7n1	41	32	true	0.638	true	2.131
state space48	48	48	true	0.387	true	41.688
PWM extended	49	31	true	2.510	true	12.776
lnet p8n1	51	40	false	1.080	false	2.678
cn_crr7	56	15	n/a	n/a	true	3.862
cn-crr10	80	21	n/a	n/a	true	11.128
cn-crr15	120	31	n/a	n/a	true	45.246
cn crr25	200	51	n/a	n/a	true	294.228

ACRONYMS

ACN	=	Asymmetric Choice Net
BDD	=	Binary Decision Diagram
BRG	=	Basis Reachability Graph
CAD	=	Computer-Aided Design
CPS	=	Cyber-Physical System
EFC	=	Extended Free-Choice net
FCN	=	Free-Choice Net
FMS	=	Flexible Manufacturing System
FPGA	=	Field-Programmable Gate Array
FSM	=	Finite State Machine
GPU	=	Graphics Processing Unit
ILPP	=	Integer Linear Programming Problem
IOPT	=	Input-Output Place-Transition net
IoT	=	Internet of Things
LTL	=	Linear-time Temporal Logic
MG	=	Marked Graph
MIP	=	Mixed Integer Programming

Marcin Wojnakowski

NP-hard	=	Non-deterministic Polynomial-time hardness
PLC	=	Programmable Logic Controller
PN	=	Petri Net
PNH	=	Petri Net Hippo format
PNML	=	Petri Net Markup Language
RAM	=	Random Access Memory
SM	=	State Machine
SMC	=	State Machine Component
UCF	=	User Constraint File
UML	=	Unified Modeling Language
VHDL	=	Very high-speed integrated circuit Hardware Description Language
XML	=	Extensible Markup Language

SUBJECT INDEX

A

Acceleration, exponential 43
Accuracy 62, 82
ACN (asymmetric choice net) 24, 26, 27
Algebra 14, 44, 49, 80, 84, 92
  linear 14, 49, 80, 84, 92
Algorithm 10, 11, 30, 39, 50, 51, 52, 82, 84, 85, 87, 89, 92, 93
  advanced 50
  available 10
  classic 87
  computer 11
  coverage-based 84
  efficient 39
  form-based 89
  graph-based 84, 85, 92
  linear algebra-based 87, 93
  parallel 51
  polynomial time 51
  selected 82
  slow analysis 30
  space-based 85
  systematic 52
Analytical method, perfect 83
Analytical problem 4
Application 20, 63, 82, 91
  key 91
  optimal 82
  perfect 20
  special 63
Arcs 15, 17, 20
Array mathematics 74
Asymmetric choice net (ACN) 24, 26, 27
Automatic emergency 2
Axe 46
  available 46
  new 46

B

Behavior 25, 37, 49, 63
  binary 49
  decision 25
  logical 37
  unpredictable 63
Benchmarks 15, 91
  multi robot 15
  selected 91
Binary decision diagrams (BDD) 39
Block diagrams 13
Board game 85, 87, 90, 92, 93
Boolean 43
  satisfiability 43
  variable 43
Boundedness 83, 84, 89, 91
  unresolved 89
  algorithms 83
  experiments 91
  verification 84
Boundedness analysis 46, 48, 62, 64, 65, 67, 78, 86, 87, 88, 89, 102
  cover-based 78
  efficient 102
  preliminary 89
  procedure 86

C

Cartesian product 14
CNC machine 92, 93
Comfortable prototyping framework 14
Complexity 4, 10, 11, 12, 46, 49, 52, 53, 73, 86, 88
  analytical 49
  exponential 4, 12
  computational 10, 11, 12, 46, 49, 52, 53, 73, 86, 88
Computational problems 10, 11, 43
  simple 11
  well-specified 10

Computer-aided 1, 38, 40, 83
design (CAD) 1, 40, 83
system 83
tools 38
Computer science 12, 13, 43
theoretical 43
algorithms 12
area 13
Concurrency 13, 25, 52, 84
efficient 52
flow diagrams 13
Concurrent control 4, 14, 38, 39, 62
designers 62
specification 38
system modeling 14
specifications 4
structures 39
Construction 39, 52, 84
full-graph 84
simple 52
techniques 39
Control systems 1, 2, 3, 4, 5, 47, 49, 53, 82, 83, 95, 97, 99, 101
dedicated 2
designed concurrent 3, 47
modern concurrent 3
net-based concurrent 49, 53
simple 1
Coverability 37, 38, 40, 41, 46, 48, 50, 53, 62, 64, 65, 75, 76, 77
analysis 53
graph 37, 38, 40, 41, 46, 48, 50, 62, 64, 65, 75, 76, 77
Coverage 66, 67, 80, 87, 93, 100
additional 93
boundedness analysis based on 66, 67
plafce invariants 87
potential 100
reject place-invariant 80
CrossroadSM FPGA 85, 87, 88, 90
Cyber-physical systems (CPS) 1, 10, 13

D

Deadlock 39, 47
analysis 47
markings 39
recovery policy 47
Design 14, 52, 74, 95
path 14
processes 74, 95
solutions 52
Diagnosability analysis technique 49
Discrete event system 52

E

Emergency 2, 3
brake 3
signal 2
Entries 23, 28, 42, 71
nonzero 23, 42, 71
zero 28, 71
Equation 4, 13, 30, 42, 51
linear 4, 30
linear programming 51
mathematical 13
solving 42
Experimental 5, 53, 64, 66, 70, 72, 82, 84, 85, 87, 88, 89, 91, 93, 94, 95, 102
evaluation 53
research 53, 82, 84, 88, 91, 93, 102
results 5, 64, 66, 70, 94
verification 72, 82, 85, 87, 89, 91, 93, 95

F

Finite graphs 38, 62
Flexible manufacturing systems (FMS) 1
Formal 3, 14, 25
description 3, 14, 25
mechanisms 3
Fourier-Morzkin method 43
FPGA(s) 37, 84, 89, 95
devices 95
implementation 84

G

Gauss-Jordan 27, 69, 70, 71, 73, 88, 101
elimination 27, 69, 70, 71, 73, 88, 101
implementation 73
Gaussian Elimination 73
Graph 12, 40, 46, 49, 64, 83, 85, 88
construction 46, 64
finite reachability 40
reduced reachability 85, 88
reference reachability 40
representation 46
theory 12, 83

traditional reachability 49
Graphical 10, 14, 95
editor 14
projection 95
specification 10

H

Hippo 82, 83, 84, 95, 96, 101, 102
library 101
project 82, 83, 95, 96
system 82, 83, 84, 102
tools 83
HiPS tool 51

I

Implementation 37, 96
hardware 37
team 96
Input 1, 2, 11, 12, 15, 16, 17, 20, 49, 50, 53, 97, 100
additional 20
logical 50
data 2, 49, 53
places 15, 17
set 11, 12
signals 20, 97, 100
transitions 16
values 1
Integer 30, 53
arithmetic 53
vectors 30
Internet of Things (IoT) 1
Invariant(s) 4, 28, 38, 43, 44, 50, 53, 65, 66, 76, 79, 80, 84, 87, 88
computation 28, 80, 87
computed 4
computing 38, 44, 84
coverage 65, 66, 88
form 50
inductive 53
minimum 80
new 66
non-minimal 43, 44
obtained 66, 76, 79
reduced 87
set 87

L

Library 82, 84, 92
benchmark 82
test 84, 92
Linear algebra 30, 37, 38, 43, 49, 62, 66, 75, 83, 94
approach 43, 62, 75, 83
methodology 37
technique 38, 49, 66, 94
theory 30
Linear programming 39, 51
approach 51
problems 39, 51

M

Martinez-Silva algorithm 44, 46, 50, 66, 93
classic 93
computation algorithm 66
invariants computation method 46
method 44, 50
Material 96, 100
prepared 100
processed 96
scrap 96
Mathematical 14, 95
description 95
formalism 95
notation 14
MATLAB 73, 74
Matrix 15, 27, 28, 42, 43, 44, 63, 66, 67, 69, 77, 78, 79, 82, 89, 102
application of 69
coefficient 27, 28
conjugates 44
identity 44, 77, 78, 79
incident 79
multiplication 42
operations 44
performance tests 82
reduced row 63
reduced row echelon 89
transformations 43, 102
transposition 15
union 44, 67, 78
unit 44, 66, 78
Methodologies 30, 52
efficient synthesis 52
net-based 30

Methods 10, 38, 40, 52, 62, 70, 73, 80, 86, 87, 88, 92, 95
analytical 10, 52, 95
available 92
current 86
efficient 95
exact 62
examined 87
fastest 88
fundamental 40
new 80
numerical 73
polynomial-time 70
popular 38
Microcontrollers 37
Microprocessor 95
Modeling 3, 30, 53, 95
advanced 3
language 30, 53
systems 95

N

Net 1, 10, 19, 20, 38, 41, 42, 43, 45, 46, 49, 50, 52, 53, 63, 66, 72, 76, 79, 86, 88, 92, 97
algebraic 43
bounded 86
conflict-free 52
dual 42
interpreted 1, 19, 20, 50, 97
larger 41, 88, 92
not-covered 79
property 10
reduction 38
reversible 49
safe 49, 52
small 66, 88
unbounded 45, 46, 63, 76
uncolored 53
uncovered 72
Novel 5, 10, 62, 65, 80, 82, 88, 89, 90, 94, 95, 101
developing 82
third 88
ideas 5, 62
methods 5, 10, 80, 89, 90, 94, 95, 101
technique 65
Number 4, 11, 15, 18, 27, 37, 45, 46, 49, 63
finite 4, 18, 37, 45
indefinite 63
maximal 46, 49
aximum 11
natural 15
nonzero 27

O

Output(s) 1, 15, 17, 19, 20, 26, 50, 64, 97, 100
logic 20
text 64
transitions 15, 17, 26, 97
signals 1, 19, 20, 50, 97, 100

P

Parallel optimization processes 53
Petri net 4, 14, 15, 30, 37, 38, 39, 52, 53, 62, 78, 83, 87, 88, 94
analysis methods 62
benchmarks 83
classes 52
coverage 87, 88
invariants 37
markings 62
markup language 53
modeling 14
properties 4, 38
safeness 94
specification 15
system 78
theory 30, 39
Polynomial 12, 39, 43, 52, 69, 70, 73, 80, 88, 89, 94, 97
amount 43
complexity 12
time 39, 43, 52, 69, 70, 73, 80, 88, 89, 94, 97
Product 96, 100
finished 100
produced 96
semi-finished 100
Production 96, 97
company 97
waste 96
Properties 4, 5, 10, 14, 30, 38, 39, 41, 42, 47, 95
analyzing 38
behavioral 14, 47
common 30

fundamental 10
important 4, 5, 41, 47
key 95
preservation 39
significant 42
Proposition 4, 5, 43, 69, 70, 90, 95, 101
Prototyping 14, 83

R

Reachability 39, 53
problem 39, 53
space 39
Reference algorithm 44, 62, 65, 82, 85, 91
graph-based 85, 91
Reliability 3, 10
Representation 49, 84
graphical 84
symbolic 49
Row(s) 27, 44, 51, 67, 71, 78, 79
foreach 51
lower 27
non-zero 27
nonzero 71
pairs 44, 67, 78, 79
redundant 44, 67, 78, 79
swap 71
top 71

S

Safety analysis 50
Set 11, 12, 13, 39, 77, 88, 93
empty 11, 77
finite 12, 13
output 11
reachability 39
reduced 88, 93
Siphon analysis 39
Solutions 1, 30, 62, 63, 64, 66, 70, 76, 82, 91, 101, 102
available 82
classic 91
comprehensive 102
novel 63, 64, 66, 70, 76
perfect 62
popular 101
smart 1
trivial 30
Synthesis algorithm 49, 52
novel 52
System(s) 1, 10, 13, 14, 15, 20, 30, 39, 49, 52, 63, 66, 95
advanced 1
cyber-physical 1, 10, 13, 52
designers 10
discrete-event 49
distributed 10
homogeneous 30
modeled 49, 95
models 14
mulit-robot 66
operation 20
projected 10
properties analysis 30
reachability analysis 39
real-life 15
states 63

T

Theorems 23, 43, 52
Theory 14, 50, 52, 83
base 14
basic 50
hypergraph 52, 83
Time 11, 51, 75, 82
complexity 11, 82
efficiency 75
polynomial 51
Tokens 15, 17, 18, 37, 40, 45, 46, 49, 62, 63, 64, 75, 76, 78, 79
Tools 30, 46, 53, 62, 73, 83, 97
analytic 46
current 62
mathematical 73
model verification 53
popular CAD 97
standalone 83
strict verification 30
Toudic method algorithm 43
Transition-disjoint cycles 49
Transitions 17, 22, 23, 39, 64, 67, 86
fired 64
firing 17, 22, 23
lnet 86
single 39
third 67

V

Valmari's set 39
Values 11, 51, 66
  absolute 11
  optimal 51
  zero 66
Vector 42, 43, 84
  candidate 43
  foreach 42
  initial marking 84